AERO BALANCE
AND AERO MAP
GENERATION
AND USEFUL

It is the first book in a large and special series of books, dedicated to motorsport in general; it will cover aerodynamics, suspension, engines, dynamics, etc. ..... Everything you need to learn how to design a full car.

The aim of this series is also to say that I would like to teach again in a university.

I hope that this series will be a success and that I will be able to transmit all my knowledge and all my experience.

@TimoteoBriet

Center of Pressure (CP)

CENTER OF GRAVITY (CG)

The center of pressure is the point where the total sum of a pressure field acts on a body, causing a force to act through that point. The total force vector acting at the center of pressure is the value of the integrated vectorial pressure field. The resultant force and center of pressure location produce equivalent force and moment on the body as the original pressure field. Its position in respect the center of gravity define the dynamics of the car. Let's see the 3D calculation of CG (the most important is "x"). The location of the center of gravity is vital if we want to understand and predict certain behaviors of the car. For example when we brake, we are transferring load from back to the front tires, depending on the location of the "CG". Therefore the higher you are, the more load is transferred for a given deceleration. The same happens when we apply power.

Basicly:

- If CG is forward CP, the car have understeer.

- If CG is behind CP, the car have oversteer.

Downforce Balance

High Front to Rear Downforce-

High Rear to Front Downforce-

High speed oversteer.

High speed understeer.

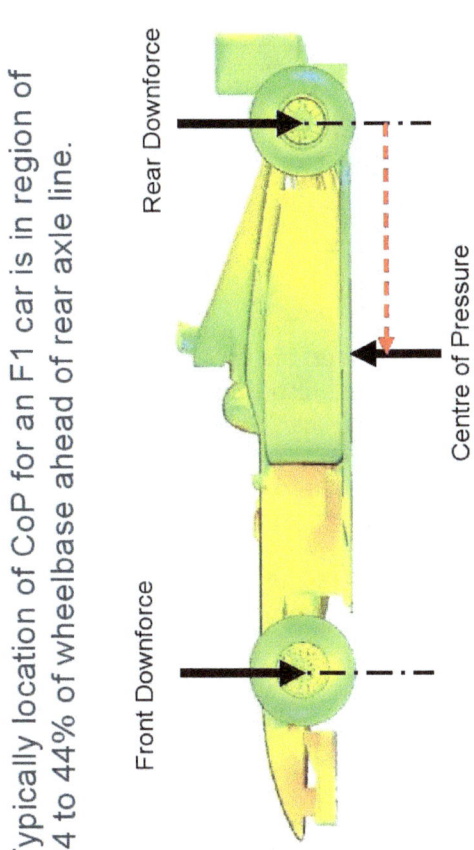

Typically location of CoP for an F1 car is in region of 34 to 44% of wheelbase ahead of rear axle line.

Rear Downforce

Centre of Pressure

Front Downforce

This concept is very important in order to know the car aero – behavior. That is:

Wind

○ Pressure center

To carry out the calculation, previously we must know the weight distribution on each wheel and basic dimensions of the car. The procedure is very simple and has three main steps.

CG. Longitudinal position:

We calculate the moment about the rear axle, we apply the following formula:

$$b = \frac{W_f \cdot L}{W}$$

Wf: Front axle weight.

W: Total weight.

X0 axis: Longitudinal axis passing through front and rear wheels' centers.

Z0 axis: Vertical axis through CG.

If CG is at a low height, the total lateral weight transfer will be smaller. This, combined that the coefficient of friction decreases after a certain point of vertical load, makes that the optimum CG has low height. If we add more vertical load to the tire, grip will decrease and the cornering speed will decrease too.

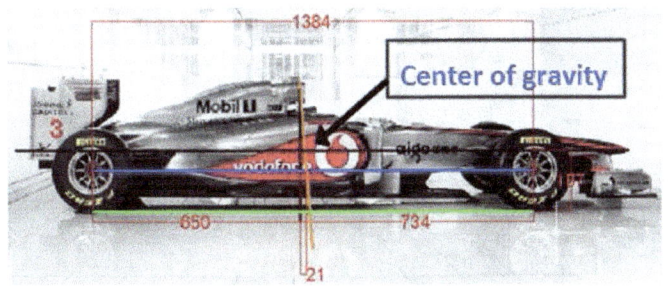

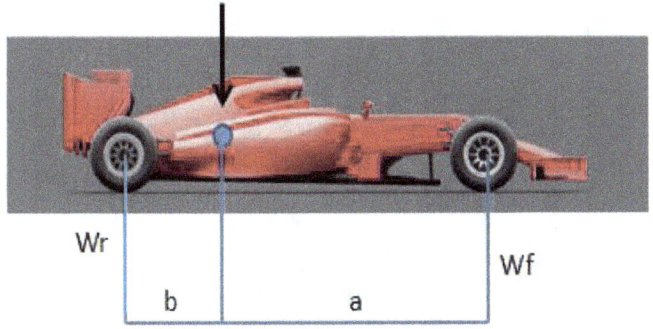

The calculation of the moment is:

$$h = \frac{(W_f \cdot L) - (W \cdot b)}{W \cdot \tan \theta}$$

Then: $hg = r + h$

r : Wheel radius

hg : "CG" vertical height.

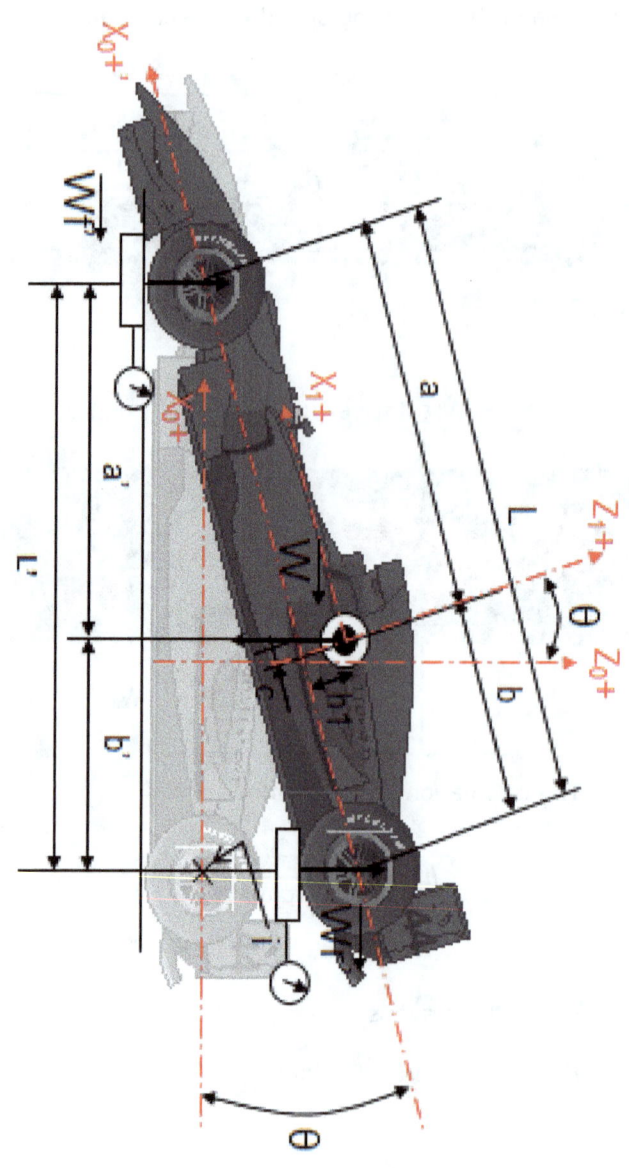

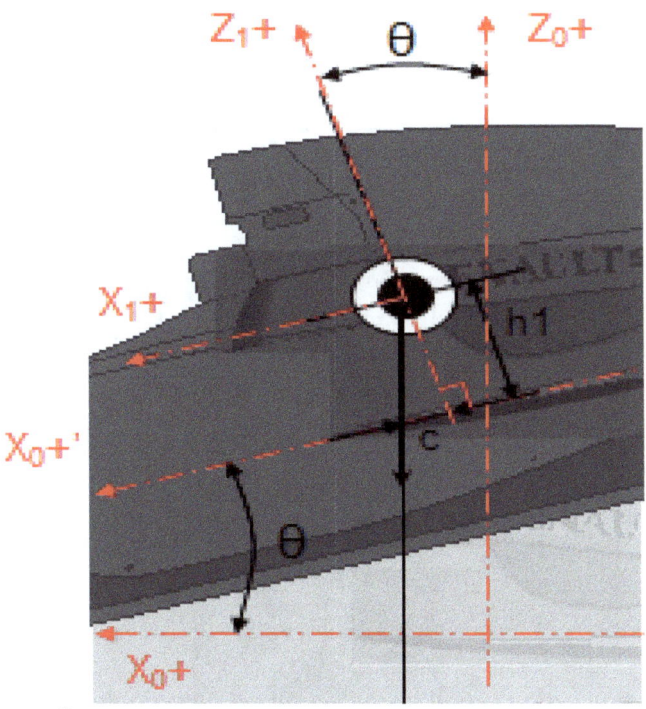

Tf: Front track.

Tr: Rear track.

Wfl: Front left weight.

Wfr: Front right weight.

Wrl: Rear left weight.

Wrr: Rear right weight.

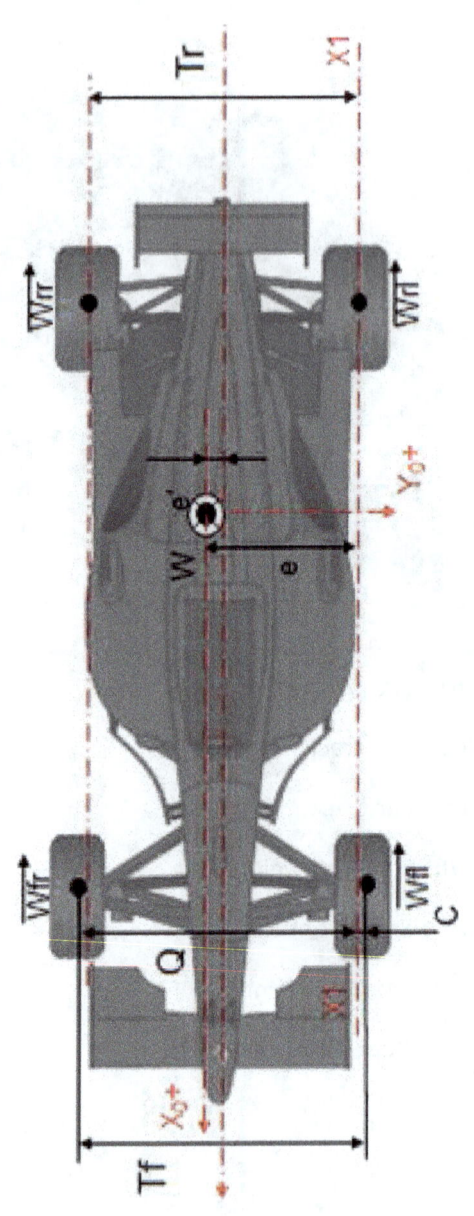

$$Wf = Wfl + Wfr$$

$$Wr = Wrl + Wrr$$

$$C = (Tf - Tr) / 2$$

$$Q = Tf - C$$

$$e' = e - (Tr / 2)$$

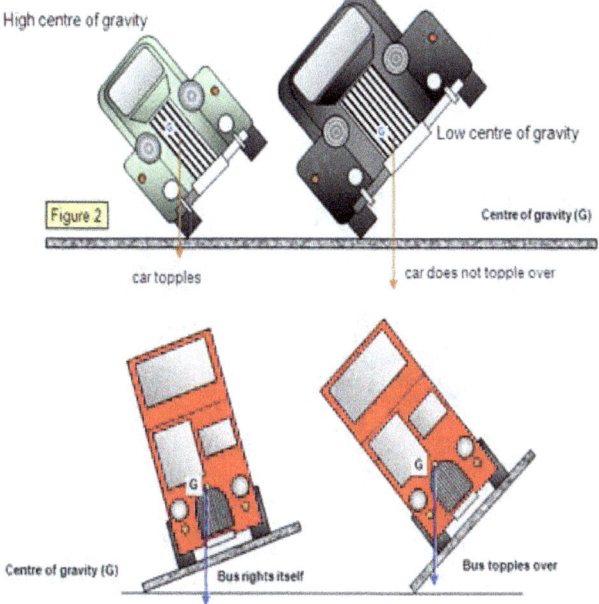

High centre of gravity

Low centre of gravity

Figure 2

Centre of gravity (G)

car topples

car does not topple over

Centre of gravity (G)

Bus rights itself

Bus topples over

Dynamic of gravity center:

If the gravity center is low height, the weight transfer will be lower; we will see that, in Chapter 15.

One think very important is the CG position and the inertia moment full:

Is a reduced moment of inertia in a car is always better?

I was having a talk with a racecar engineer about ballast (weight) that can be moved to setup weight distribution in a race car; he told me that you must always have to try the minimum moment of inertia.

- **Effects of polar moments of inertia**

Here is a example of a V8 engine with a typical transmission, Packaged into a sports car.

$$\Sigma M(I_0) = W_{Eng}(d_{Eng})^2 + W_{Tran}(d_{Tran})^2$$
$$I_0) = 600lb(40in)^2 + 240lb(10in)^2$$
$$= 984{,}000lb \cdot in^2$$

Example:

$W_{Eng} = 600 \ lb$

$W_{Tran} = 240 \ lb$

$d_{Eng} = 40 \ in$

$d_{Tran} = 10 \ in$

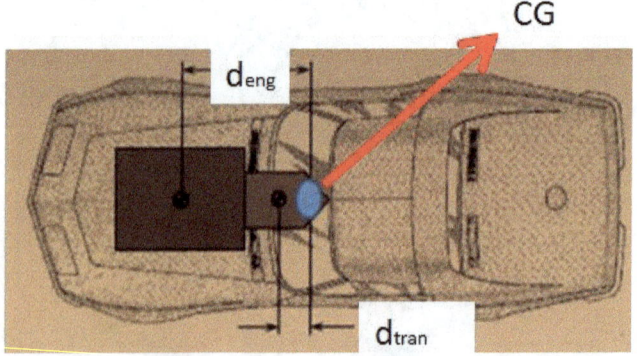

But, I think that that may be the case in a track with many chicanes where you need a fast yaw response (high speed rotational), but maybe in other tracks with open fast curves you may want a car that is less prone to a spin.

Is complicate the question: The Dynamic Index.

This number, mean that the mass distribution in a car, is the best; a good repartition of masses; for obtaining this number; sample:

B	C	D		
Wheelbase, L	2,4	(m)	L	
offset, X	0,2	(m)	X	
front axle to CG, a	1,18842105	(m)	a	
rear axle to CG, b	1,21157895	(m)	b	
inercia total, Izz	1194,4 (kgm^2)		Izz	
massa, m	950 (kg)		m_total	
weight %front	50,5 (%)			
D.I.	0,87321188			

component	mass, m (kg)	Inertia, I (kgm^2)	position to datum (m)	position to CG (m)	mK^2+I (kgm^2)	position_to_front_axle (m)	k_component check
engine+gearbox	200	12	0,5	0,888421053	169,858393	0,3	60
chassis	100	90	1,5	-0,111578947	91,2449861	1,3	130
battery	10	0,07	2,7	-1,311578947	17,2723934	2,5	25
fuel tank+50%fuel	30	2	2,6	-1,211578947	46,0377064	2,4	72
exhaust+cat	10	3,5	2,7	-1,311578947	20,7023934	2,5	25
seats	20	2	2,2	-0,811578947	15,1732078	2	40
driver	80	7	2,2	-0,811578947	59,692831	2	160
bits'n bobs	50	35	1,5	-0,111578947	35,6224931	1,3	65
front susp+tyres	80	20	0,2	1,188421053	132,987568	0	0
rear susp+tyres	80	20	2,6	-1,211578947	137,433884	2,4	192
ballast F	30	2	0	1,388421053	59,8313906	-0,2	-6
ballast R	30	2	2	-0,611578947	13,2208643	1,8	54
Diff	30	1,3	2,6	-1,211578947	45,3377064	2,4	72
car body	200	350	1,4	-0,011578947	350,026814	1,2	240
total	950				1194,44263		1129

	B	C	D	E	F	G
1	Wheelbase, L	2,4	(m)	L		
2	offset, X	0,2	(m)	X		
3	front axle to CG, a	1,18842105	(m)	a	I25/m_total	
4	rear axle to CG, b	1,21157895	(m)	b	L-a	
5	inercia total, Izz	1194,4	(kgm^2)	Izz	G25	
6	massa, m	950	(kg)	m_total	C25	
7	weight %front	50,5	(%)		b/L*100	
8	D.I.	0,87321188	-		(Izz/m_total)/(a*b)	

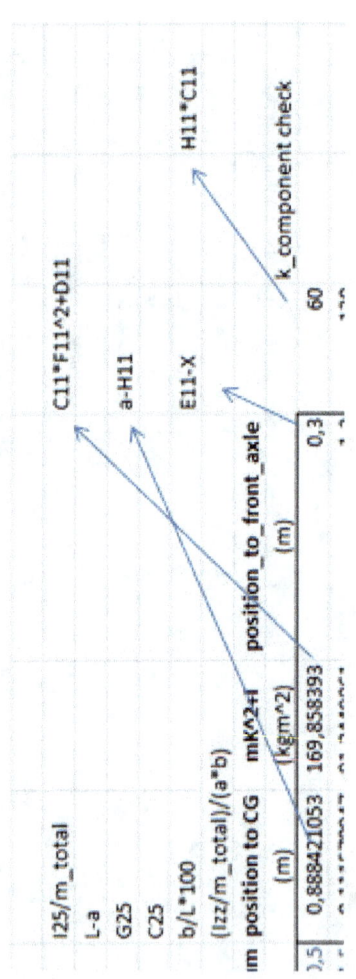

The valor ideal, for sport car (race cars) is DI = 0.8 – 0.9. That is complicate to know it, because the main goal, is to have a yaw acceleration very fast; that is:

Is better so, calculation "PI" i order to have the better yaw acceleration.

CENTER OF PRESSURE (CP)

Note: not is the same PRESSURE CENTER than AERODYNAMIC BALANCE POINT.

Pressure Center or AERODYNAMIC CENTER, against AERODYNAMIC BALANCE POINT:

- AERODYNAMIC BALANCE POINT: work also with MOMENTS. For example, the DRAG, produce PITCH MOMENT and this effect, is necessary to know. Very important for RACE CARS.

- AERODYNAMIC CENTER or PRESSURE CENTER: not work with moments. Not important for RACE CARS.

➔ But normally, is common say PRESSURE CENTER as AERODYNAMIC BALANCE POINT.... But is necessary to know that is wrong....

As we have a center of aerodynamic pressure to a wing or a small piece, we also have the center of pressure of a complete car. This point is absolutely necessary to know and determine precisely: is the "balance" of the car ("x" axis, principally; the same process for "y" and "z" axis).

If you have or have access to a wind tunnel, we can easily know the position of the center of pressure. In competition and dynamically speaking, we need to know the longitudinal position of the "CP", that is, the position of the "CP" between the axes wheel. The vertical position or height of the "CP" matters to us, but less. This is because what interests us most is to know how the downforce is distributed between the front and rear axle. In order to determine the CP position, assuming "FF" is the front downforce, and "FR" is the rear downforce, the percentage of front downforce over total downforce is:

$$\%Front = \frac{FF}{FF + FR} * 100$$

Multiplying by wheelbase, we obtain the longitudinal position.

The positions themselves of the "CG" and "CP" as well as their relative positions, define the dynamics in track. If the pressure center is displaced from the center of gravity moments occur, making the vehicle unstable. If "CP" is ahead regarding the "CG" we can have understeer if the downforce is scarce. If it is delayed with respect to the "CG" oversteer occurs. If a crosswind appears and the "CP" and "CG" are shifted, depending on the wind direction, it could increase the possibility of oversteer or understeer occurs.

Method 1:

Let the next nomenclature and notation:

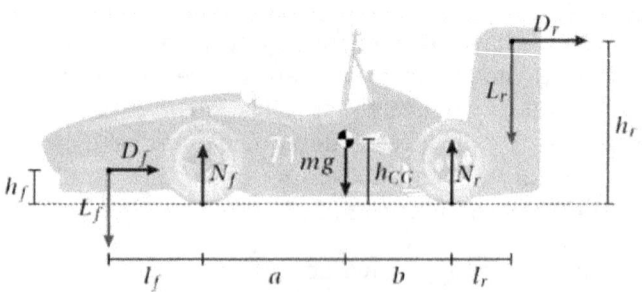

The force in rear contact patch is (the Drag also produce moment):

$$N_r = \frac{amg + (l_r + a + b)L_r - l_f L_f + (h_r - h_{CG})D_r + (h_f - h_{CG})D_f}{a + b}$$

First calculate Fx, Fy and My0; My0 is the moment about origin (axle front wheel).

After:

$$M_y = M_{y0} + dxF_z - dzF_x$$

Where dx and dz are the distances from the origin to the contact patch parallel and normal to the base of the car, L (wheelbase):

$$F_{front} = M_y / L$$

→ Suppose a school plane to learn to fly: the student must have a docile and quiet plane to absorb the mistakes committed and especially lazy to the changes of course made by the student. In short, a not dangerous plane. This plane should have the center of pressure above the center of gravity. This is the reason why the school aircraft have the wings at the top. It is said in this case, the plane is in stable equilibrium.

Now imagine an aerobatic plane: has the opposite characteristic of the above-mentioned for school plane:

- It is faster to the changes in course of the pilot.

- It is quick to execute maneuvers requested.

- Doesn't matter to fly upwards or downwards.

The plane therefore should have the center of pressure as close to the center of gravity as possible. This is the reason why acrobatic aircraft have wings in the middle of the fuselage:

With cars, it is exactly the same: If the center of pressure was at the same point of the center of gravity, the car wouldn't understeer nor oversteer.

Method 2:

Mathematically, we can define the center of pressure:

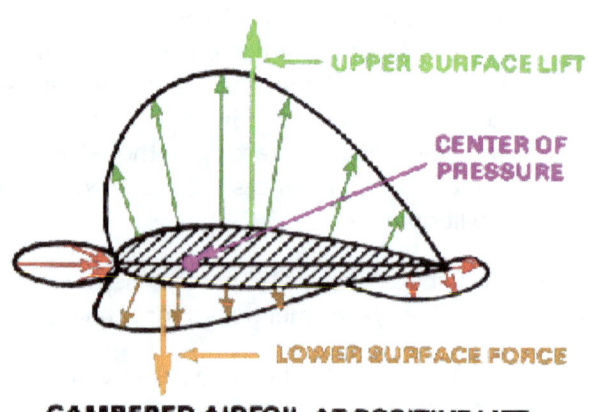

CAMBERED AIRFOIL AT POSITIVE LIFT

$$x_{CP} = -\frac{M_o}{L} \approx \frac{\int_{x=0}^{x=c}(p_{ext} - p_{int}) \cdot x \cdot ds}{\int_{x=0}^{x=c}(p_{ext} - p_{int}) \cdot ds}$$

Method 3:

If we show the pressure against the "x" position:

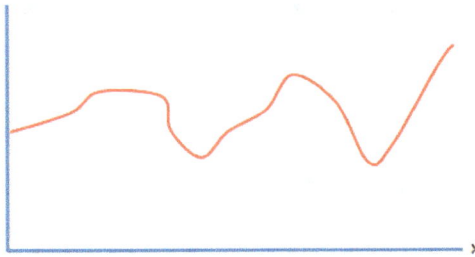

and calculate the center of this graphic in order to calculate the pressure center in "x", is necessary to representing the pressure in "z" axis. Another possibility is to representing the pressure multiplying by the "z" component of area (element mesh with this pressure).

If we representing pressure absolute, is not possible to calculate the center because we not know if the pressure value is to up or to down.

Method 4:

This new method is ideal for simulating in CFD.

Is based in: If we rotate freely in pitch the car, and the car is not moving, the point of action of the rotation axis is the CP. That is: from CFD, we fix the axis of rotation of the front axle drive and analyze the rotational movement of the car (if not exist purpose for example); calculating the moment, we know the CP:

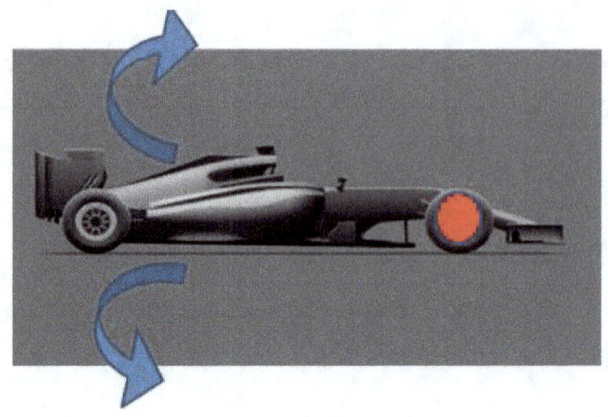

Method 5:

$$cdp(x-axis) = \frac{\sum Area(z-axis) * \Pr essure * Position(x-axis)}{\sum \Pr essure * Area(z-axis)}$$

Method 6:

For example, from Star CCM as CFD code, is possible to calcul from .java file:

```java
// STAR-CCM+ macro: centropresion.java

package macro;

import java.util.*;

import star.common.*;

import star.base.neo.*;

import star.base.report.*;

public class centropresion extends StarMacro {

  public void execute() {

    Simulation simulation_0 =
```

```
getActiveSimulation();

UserFieldFunction userFieldFunction_0 =

simulation_0.getFieldFunctionManager().createFieldFunct
ion();

    userFieldFunction_0.setPresentationName("Px");

userFieldFunction_0.setDefinition("$$Position[0]*$Pressu
re");

    UserFieldFunction userFieldFunction_1 =

simulation_0.getFieldFunctionManager().createFieldFunct
ion();

    userFieldFunction_1.setPresentationName("Py");

userFieldFunction_1.setDefinition("$$Position[1]*$Pressu
re");

    UserFieldFunction userFieldFunction_2 =

simulation_0.getFieldFunctionManager().createFieldFunct
ion();

    userFieldFunction_2.setPresentationName("Pz");

userFieldFunction_2.setDefinition("$$Position[2]*$Pressu
re");

    SurfaceIntegralReport surfaceIntegralReport_0 =

simulation_0.getReportManager().createReport(SurfaceI
ntegralReport.class);
```

```java
surfaceIntegralReport_0.setScalar(userFieldFunction_0);

    surfaceIntegralReport_0.setPresentationName("Px");

    SurfaceIntegralReport surfaceIntegralReport_1 =

simulation_0.getReportManager().createReport(SurfaceI
ntegralReport.class);

    surfaceIntegralReport_1.setPresentationName("Py");

surfaceIntegralReport_1.setScalar(userFieldFunction_1);

SurfaceIntegralReport surfaceIntegralReport_2 =

simulation_0.getReportManager().createReport(SurfaceI
ntegralReport.class);

    surfaceIntegralReport_2.setPresentationName("Pz");

    SurfaceIntegralReport surfaceIntegralReport_3 =

simulation_0.getReportManager().createReport(SurfaceI
ntegralReport.class);

surfaceIntegralReport_2.setScalar(userFieldFunction_2);

PrimitiveFieldFunction primitiveFieldFunction_0 =

    ((PrimitiveFieldFunction)
simulation_0.getFieldFunctionManager().getFunction("Pre
ssure"));

surfaceIntegralReport_3.setScalar(primitiveFieldFunction
_0);
```

```java
surfaceIntegralReport_3.setPresentationName("P");

ExpressionReport expressionReport_0 =

simulation_0.getReportManager().createReport(Expressio
nReport.class);

expressionReport_0.setPresentationName("CP x");

expressionReport_0.setDefinition("$PxReport/$PReport")
;

ExpressionReport expressionReport_1 =

simulation_0.getReportManager().createReport(Expressio
nReport.class);

expressionReport_1.setPresentationName("CP y");

expressionReport_1.setDefinition("$PyReport/$PReport")
;

ExpressionReport expressionReport_2 =

simulation_0.getReportManager().createReport(Expressio
nReport.class);

expressionReport_2.setPresentationName("CP z");

expressionReport_2.setDefinition("$PzReport/$PReport");
  }

}
```

In this file in java, there is not the viscous forces, but the error is very little.

Method 7: let these 2 designs:

If we calculate the CDP from the expression: Moment Pitch / Down Force:

- Case Photo 1 : true.

- Case 2 Photo 2 : wrong.

Why ?

Because in the second case, we not work with the pitch moment created by the Drag. So is necessary another expression for Case two, or in general in race cars:

Distance from front axis = (Pitch Moment – (Drag * HeightCDP)/Downforce)

This Drag Moment Pitch is generated, normally and basically, by the rear wing.

Method 8:

The Center of Pressure (COP) of a body is defined as a point on the body about which the net moment is always zero, be it in any direction.

Finding the center of pressure in 3D is not a straightforward as in 2D, since what you end up solving for is the equation of a line about which the net force acts, instead of the point which is obtained in a 2D case.

For 3D, the following image illustrates the geometry that we are solving on, where Ft=net force vector, Mt=net moment vector, and rs=vector position of the COP.

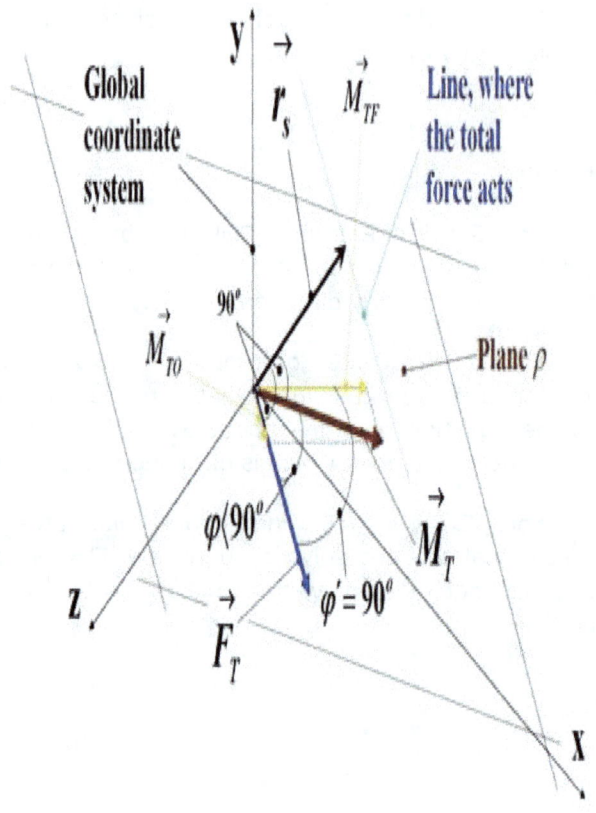

$$x = \frac{\int x\,dF_y}{F_y}, \quad y = \frac{\int y\,dF_x}{F_x}$$

➔ Example with CFD tool: Front wing F1 rules 2017:

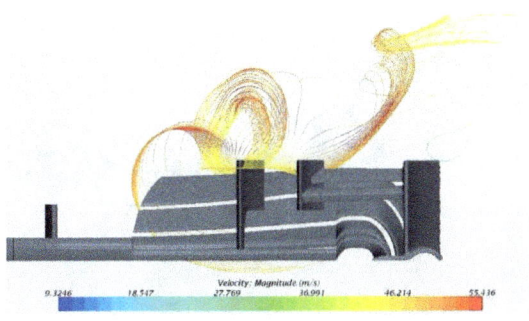

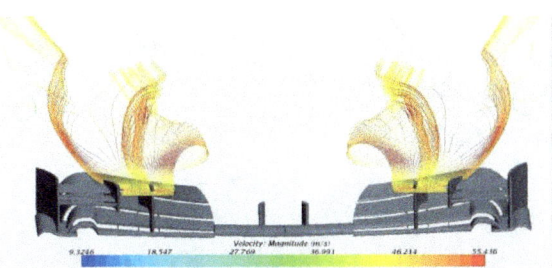

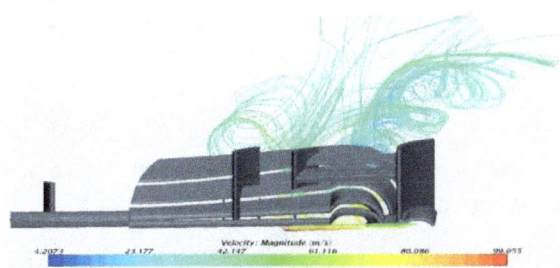

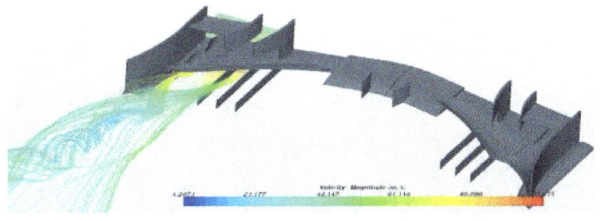

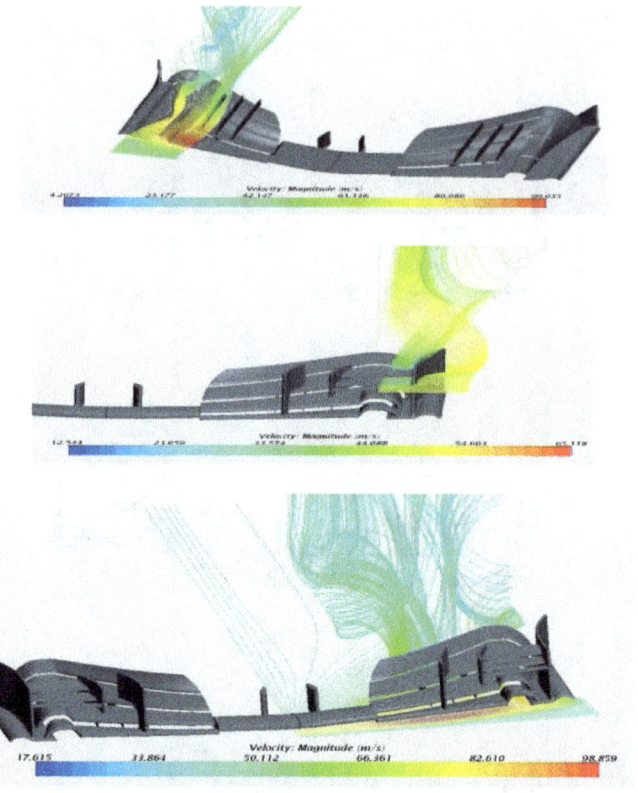

The pressure center is located in (placed in chassis-wing contact….):

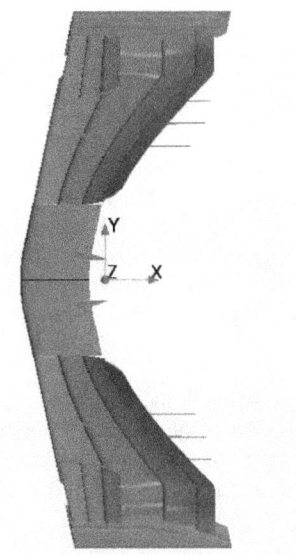

Method 9:

Is possible to calculate the CDP position in 3 axis, from CFD simulation; that is:

The full car is free so that it moves in any direction. In this way, it is possible to calculate the force in each contact path. There are CFD codes that automatically, it´s possible to calculate the CoP; but calculate it by moment / downforce, it´s enough....

Method 10: calculating the moment in each axis: must to be "0".

TOP SPEED IN CORNER; AERO EFFECT

The problem that we face is to be able to calculate the maximum speed cornering of a vehicle taking into account the effect of the downforce and especially its position (CP - balance), having to work together, knowledge of dynamics and aerodynamics.

For starters we plan to perform the most complex model that we can use for this purpose using the rigidities of springs and wheels for maximum speed and determine whether the car will tend more to oversteer or understeer.

in this model the weight transfer is contemplated coming to calculate the vertical force at each wheel so that this force multiplied by the coefficient of friction of our tire, we will have the maximum force that can generate a tire and adding both tires each axis will have the strength in each axis.

For this calculation will study the force in each axis and not on each tire and we will take the simplification that the coefficient of friction is constant across the range of vertical load of the tire, of course that's not entirely true.

It can be shown that the transfer of weights on an only curve does not affect the vertical load on the whole of each axis and therefore, if we consider the simplification of a constant coefficient of friction, the lateral force not altered. This is given as:

$$F_{ext\_front} = F_{static}$$
$$+ Transferencia_{mass\ sprung\ spring}$$
$$+ Transferencia_{mass\ sprung\ by\ RC}$$
$$+ Transferencia_{mass\ unsprung}$$

$$F_{interior\_delantera}$$
$$= F_{estática}$$
$$- Transferencia_{masa\ susp\ por\ muelles}$$
$$- Transferencia_{masa\ susp\ por\ RC}$$
$$- Transferencia_{masa\ no\ susp}$$

So the total vertical force axis coincides with the static weight bearing on that axis. This simplifies things and thus the maximum allowed speed on the front axle is a simple balance of forces on each axis:

$$\mu F_{f\_estático} = m_f \frac{v^2}{R} \rightarrow v = \sqrt{\frac{\mu \cdot F_f \cdot R}{m_f}}$$

The same for the rear axis.

<u>Front Axis:</u>

Exterior wheel $Fz0f = mg\dfrac{b}{2*Wb} + mg\dfrac{h}{Tf}*\dfrac{Kf}{Kt} +$

$\dfrac{0.5*Rh0*v^2*Cl*A}{2} * \dfrac{Wb-x}{Wb}$

Interior wheel $Fz1f = mg\dfrac{b}{2*Wb} - mg\dfrac{h}{Tf}*$

$\dfrac{Kf}{Kt} + \dfrac{0.5*Rh0*v^2*Cl*A}{2} * \dfrac{Wb-x}{Wb}$

m= peso del vehículo
g=aceleración de la gravedad
b=distancia del centro de gravedad al eje trasero
Wb=batalla del vehículo
h=altura del centro de gravedad
Tf=anchura del eje delantero
Kf/Kt=reparto de rigidez en el eje delantero
Rh0=valor de la densidad del aire
v=velocidad del paso por curva
Cl=coeficiente aerodinámico del coche
A=área frontal de ataque aerodinámico
x=posición del centro de presiones

<u>Rear axis:</u>

Exterior wheel $Fz0r = mg\dfrac{a}{2*Wb} + mg\dfrac{h}{Tr}*$

$\dfrac{Kr}{Kt} + \dfrac{0.5*Rh0*v^2*Cl*A}{2} * \dfrac{x}{Wb}$

Interior wheel $Fz1r = mg\dfrac{a}{2*Wb} - mg\dfrac{h}{Tr}*$

$\dfrac{Kr}{Kt} + \dfrac{0.5*Rh0*v^2*Cl*A}{2} * \dfrac{x}{Wb}$

a=distancia del centro de gravedad al eje delantero
Tr=anchura del eje trasero
Kr/Kt=reparto de rigidez en el eje trasero

Force available in every axis:

The force on each axis is the sum of the forces available at each wheel. Finally the total available power on each axis is the sum of the available wheel multiplied by the friction coefficient of the tire, which in this case has been replaced by two coefficients Pacejka. With this, we have the following.

$$FzF = Fz0f + Fz1f$$
$$Fdisponible = a1 * FzF^2 + a0 * FzF$$

The calculation for the rear axle is analogous, changing the values front the rear.

If we include also introduce downforce coefficient parameters downforce, aerodynamic balance (which will include as% F or percentage of front downforce) and the frontal area of the vehicle and cannot leave or variable air density according to model accuracy to seek, like modeling the tire.
Including that, the equilibrium is:

$$\mu \cdot \frac{1}{2} \cdot \rho \cdot A \cdot Cl \cdot v^2 \cdot \%F + \mu \cdot F_{f\_estático} = m_f \frac{v^2}{R}$$

$$v = \sqrt{\frac{\mu \cdot F_f}{\left(\frac{m_f}{R} - \mu \cdot \frac{1}{2} \cdot \rho \cdot A \cdot Cl \cdot \%F\right)}}$$

We obtained by using these equations, the maximum speeds in each axis according to the balance of forces so that the limiting shaft speed will be the least mark.
→ We are able so, to know the tire which lost grip.

ACCELERATION; AERO EFFECT

Another factor very important to know is the vehicle acceleration capabilities; the aero, in this case, it is not as important as aero in top speed, but is important to know it his influence.

The red values are data initials; the red and blues, are values calculated:

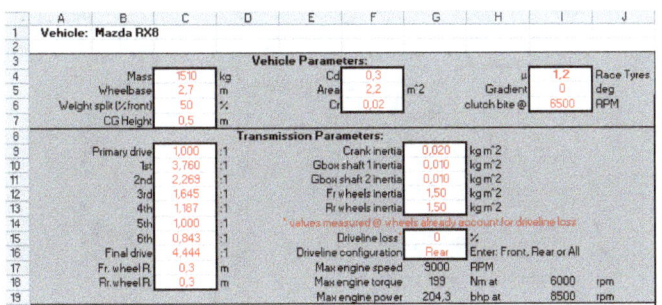

	A	B	C	D	E	F	G	H	I	J
1	Vehicle: Mazda RX8									
2										
3					Vehicle Parameters:					
4		Mass	1510	kg	Cd	0,3		μ	1,2	Race Tyres
5		Wheelbase	2,7	m	Area	2,2	m^2	Gradient	0	deg
6		Weight split (%front)	50	%	Cr	0,02		clutch bite @	6500	RPM
7		CG Height	0,5	m						
8					Transmission Parameters:					
9		Primary drive	1,000	:1	Crank inertia	0,020	kg m^2			
10		1st	3,760	:1	Gbox shaft 1 inertia	0,010	kg m^2			
11		2nd	2,269	:1	Gbox shaft 2 inertia	0,010	kg m^2			
12		3rd	1,645	:1	Fr wheels inertia	1,50	kg m^2			
13		4th	1,187	:1	Rr wheels inertia	1,50	kg m^2			
14		5th	1,000	:1	* values measured @ wheels already account for driveline loss					
15		6th	0,843	:1	Driveline loss	0	%			
16		Final drive	4,444	:1	Driveline configuration	Rear	Enter: Front, Rear or All			
17		Fr. wheel R	0,3	m	Max engine speed	9000	RPM			
18		Rr.wheel R	0,3	m	Max engine torque	199	Nm at	6000	rpm	
19					Max engine power	204,3	bhp at	8500	rpm	

Performance			
Max speed:	245	Km/h	L34
0:100Km/h:	5,76	sec	L35
0:160Km/h:	14,56	sec	L36
0-400m:	14,10	sec	L37
0-1000m:	25,92	sec	L38

This sheet is complicate but the essential is to know the importance (in this case, of the drag principally). For that, is possible run step by step; that: in every step, we know the power against the relation in gearbox; we know also the mass transfer in acceleration so we know the mass in every tyre; conclusion: we know if the tire is able to have friction in track or not; trough that, is possible to know if the tire have a contact with the track.

We have calculated already that, in chapter about lap time.

The "ideal" is giving also the engine power curves. So is necessary to know the variation power against rpm:

I	J	K	L
		Engine:	
1,2	Race Tyres	**speed**	**Torque**
0	deg	RPM	Nm
6500	RPM	2000	130
		2500	153
		3000	153
		3500	163
		4000	165
		4500	180
		5000	192
		5500	195
veline loss		6000	199
		6500	195
. Rear or All		7000	192
		7500	181
6000	rpm	8000	174
8500	rpm	8500	171
		9000	145

LAP TIME; AERO EFFECT – EXAMPLE

Obviously, the principal goal about the position "ideal" of CDP, is to reduce the lap time; see that; for example, in Montmelo circuit, we have a car:

Transmission

Primary drive	1,000	:1	PR
1st	4,3	:1	G1_
2nd	3,6	:1	G2_
3rd	3	:1	G3_
4th	2,6	:1	G4_
5th	2,3	:1	G5_
7th: 1.85 6th	2,05	:1	G6_
Final drive	3,45	:1	AR
Diff TBR	2,5	:1	TBR
Wheel radius	0,325	m	r
Driveline loss	10	%	drive_loss
Driveline configuration	RWD	Enter: FWD, RWD or 4WD	

Engine

Speed [RPM]	Torque [Nm]	Power [BHP]
3000	110	46
4000	190	107
5000	219	154
6000	237	200
7000	242	238
8000	245	276
9000	249	315
10000	260	365
11000	295	456
12000	320	540
13000	330	603
14000	347	683
15000	330	696
16000	320	720
17000	310	741
18000	300	759
19000	290	775
20000	280	787

Chassis

mass	600	[kg]
Weight split	30	[%]
CG height	0,2	[m]
Wheelbase	3,1	[m]
Track	1,425	[m]
Susp_stiff. Split	45	[%]
Brake split	40	[%]
Max brake capacity	10	[g]

Tyres

coefficients:	Front	Rear
Lat a1_	-80	-40
Lat a2_	1800	2000
Long b1_	-25	-10
Long b2_	1800	2000

Finally, about aero:

Aero

drag coeff.	0,75	-
lift coeff.	-2,25	-
Area	1,3	[m^2]
Centre of pressure X	1,8	[m]
Centre of pressure Z	0,26	[m]

With all these values, the results are:

LapTime:	
min:	sec
1	23,4

We change aero values (less downforce frontal or more rear):

Aero		
drag coeff.	0,75	-
lift coeff.	-2,25	-
Area	1,3	[m^2]
Centre of pressure X	1,9	[m]
Centre of pressure Z	0,26	[m]

LapTime:	
min: sec	
1	22,9

So, as we can see, this position is very very important ¡ ¡ ¡ ¡

PIKES PEAK – EXAMPLE IN CFD (more about, in another full chapter)

We test a first model:

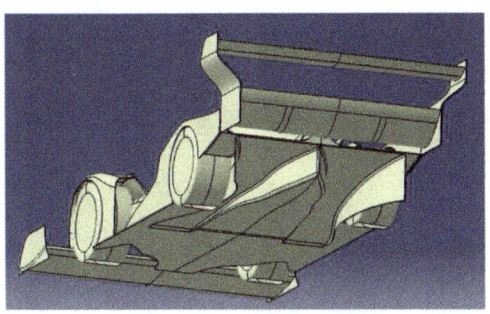

150 Km/h, ground moving and wheels rotating and 4 exhaust (258 mph speed inlet from exhausts):

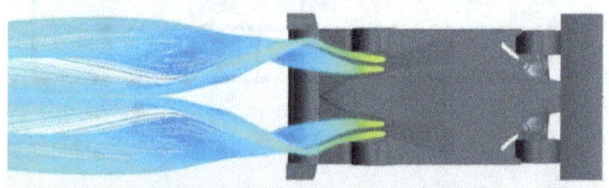

Kg	Rear Wheel	Front Wheel	Front Wing	Rear Wing	Floor	Diffuser
Downforce	3.2 (lift)	21,33 (lift)	152.85	179.6	191	207.25
Drag	4.3	15.2	9.6	75.26	0.98	31.4

The Pressure center is placed to 2.24 m from front axle.

The second model, have a variation:

Version 2:
150 Km/h, ground moving and wheels rotating:

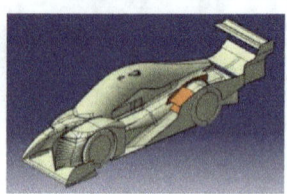

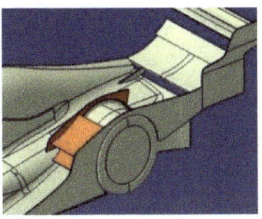

Kg	Rear Wheel	Front Wheel	Front Wing	Rear Wing	Floor	Diffuser	Full Car
Downforce	3,8 (lift)	21,2 (lift)	153.5	201.8	183.7	205.2	654.6
Drag	1.17	14.7	9.63	74.34	0.98	31.22	221.2

The Pressure center is placed to 2.23 m from front axle.

The new deflector, create in rear wing, less downforce and so, the cdp more or less, the same, although the there is more downforce in this new device….

Version 3:

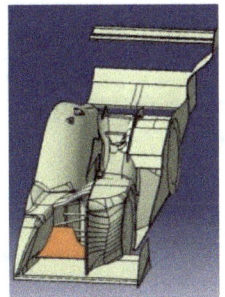

In this last case, the cdp is located to 1.99 m i i i i

Aero-Map

DESCRIPTION

At the time of installing the ideal setup on the car, we need to know how the car will respond, i.e.: what dynamic performance will have before either aerodynamic configuration. We need therefore a chart or graph that provides us that data. We need to know several variables:

- Total Downforce: FZT
- Front Downforce: FZF
- Rear Downforce: FZR
- Total Drag: FXT
- Centre Of Pressure's Position or Balance: A %

There are 2 types of aero map:

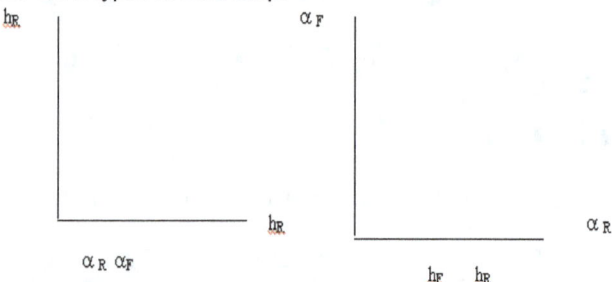

"hr" and "hf" are front and rear height.
In the first case we have, for a given aerodynamic configuration, values of forces depending on the front and rear heights.
In the second case, for a given front and rear heights, we have the same data according to each aerodynamic configuration; the vertical axis corresponds to the front angle of incidence and the horizontal, rear.
The most widely used is the first, undoubtedly.
Both front and rear heights are the heights from the intersection points between the extension of the floor of the car and each wheel axles compared to asphalt:

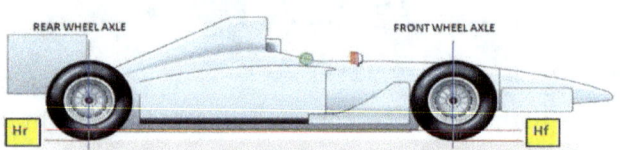

These points are imaginary, which both height measurements must be performed by "indirect" means:
This is measurement of "heights" of "accessible" points, so that such measurements correspond to the heights given by the definition of front and rear heights:
- We will measure "h" with a calibrated device:

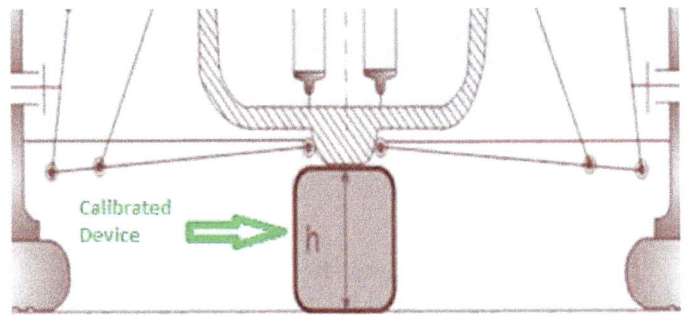

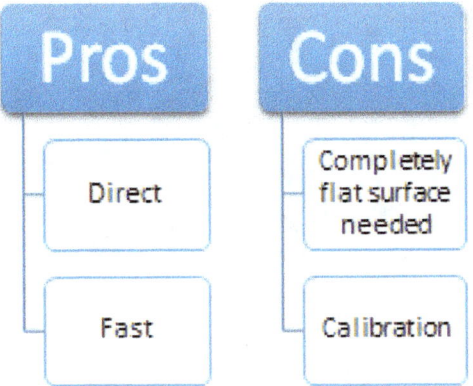

Another method to get heights is to measure other points:

- Using "setup wheels" and bar:

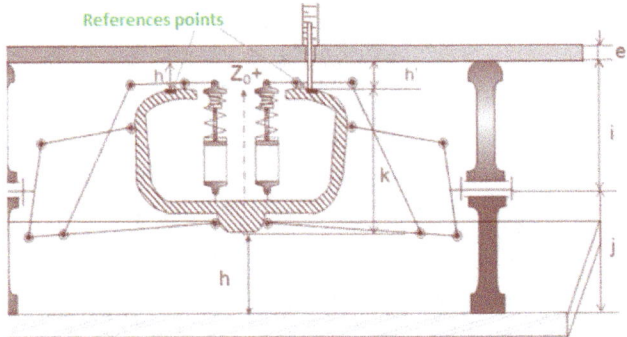

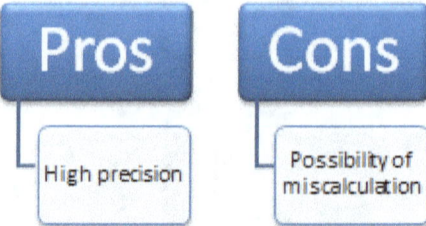

Need to know the precise value of k:

$$h = i + j + e - h' - k$$

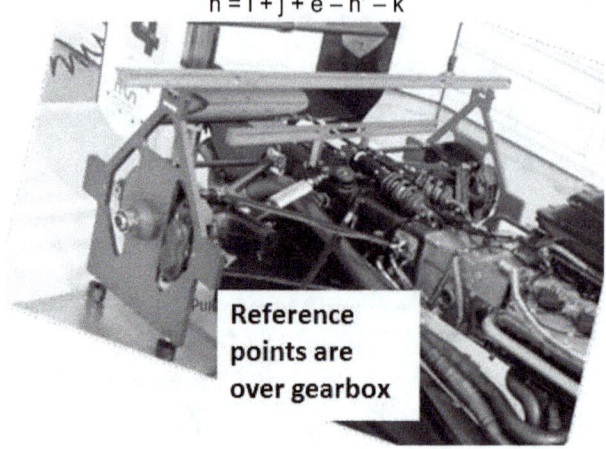

Reference points are over gearbox

Front reference points are over the nose.

An aero map is composed of different data tables:

CzT

RRH	5	10	15	20	25	30	35	40	45
FRH 5	2.697	2.730	2.807	2.840	2.865	2.872	2.848		
10	2.595	2.652	2.699	2.736	2.759	2.744	2.738	2.724	
15	2.523	2.575	2.618	2.641	2.638	2.628	2.610	2.604	2.602
20		2.501	2.542	2.545	2.541	2.534	2.527	2.505	2.493
25			2.469	2.458	2.459	2.451	2.435	2.422	
30				2.400	2.388	2.381	2.369		

CzF

RRH	5	10	15	20	25	30	35	40	45
FRH 5	1.087	1.120	1.183	1.217	1.247	1.269	1.275		
10	1.014	1.060	1.099	1.135	1.160	1.172	1.184	1.197	
15	0.958	0.998	1.033	1.061	1.073	1.083	1.092	1.103	1.118
20		0.938	0.972	0.989	1.002	1.013	1.025	1.028	1.037
25			0.916	0.927	0.944	0.955	0.961	0.967	
30				0.881	0.892	0.902	0.910		

Eff

RRH	5	10	15	20	25	30	35	40	45
FRH 5	2.877	2.983	3.079	3.110	3.130	3.131	3.111		
10	2.862	2.929	2.980	2.999	3.023	3.004	3.004	2.980	
15	2.784	2.851	2.891	2.916	2.913	2.896	2.875	2.860	2.852
20		2.776	2.816	2.811	2.810	2.795	2.787	2.762	2.742
25			2.741	2.729	2.726	2.722	2.694	2.676	
30				2.682	2.667	2.657	2.628		

The above table, corresponds to different values (particularly, total drag, front drag and efficiency) based on the front high "FRH" (vertical axis) and rear "RRH" (horizontal axis), for a specific aerodynamic configuration:

- CZT, total lift coefficient.
- CxT, total drag coefficient.
- CzF, front lift coefficient.
- CzR, rear lift coefficient.
- Eff, efficiency (lift / drag).
- Bal: % front balance (also called "A").

The basic and essential are 3: CxT, CzR, CzF, since the rest are derived from these three.
Moreover, the aero map indicates aerodynamic values depending not only from the front or rear angles, but the opening having pontoons and how we cover them, in terms of how we cover chimneys, etc....

CFG	Gills exhaust	Chimney exhaust	$\Delta C_{xT}S$	$\Delta C_{zF}S$	$\Delta C_{zR}S$
1	OPEN	OPEN	-	-	-
2	CLOSED	OPEN	-0.004	-0.005	0.012
3	OPEN	CLOSED 50%^{2}	-0.004	0.000	0.003
4	OPEN	CLOSED 100%	-0.005	-0.001	0.001
5	CLOSED	CLOSED 100%	-0.009	-0.013	0.006

Unfortunately, in the case of a Dallara GP2, the manual only provides us with the aero map of 3 aerodynamic configurations:
- Low Down-Force.
- Mid Down-Force.
- High Down-Force.

They seem few and they are, but at the end of season you realize that you may have used, in addition to those indicated in the manual, 3 configurations more...

If we work in Formula 3 Dallara does not provide any aerodynamic value: hence teams invest in aerodynamic developments to know the car: CFD techniques with previous 3D scan or wind tunnel.

GENERATION AND DISPLAY

To generate this huge amount of values, remember that a "normal" car can have 10,000 different aerodynamic configurations and more, you can use three methods:
- Calculate the parameters from CFD studies; It is therefore necessary to have the car in CAD format.
- Calculate the parameters using a wind tunnel: the quickest, most direct and precise solution; but it is the most expensive....
- Calculate the parameters using data acquired on track.

This last option is perhaps the most used by all the teams because barely have to make some investment: we have analyzed the process to follow above.
One of the main functions that possess all aero-map, is "clearly" display data. It may seem silly but it's true: there is a lot of data and need to know the areas of optimum performance of the car on a glance. We can use a color representation of each exposed value:

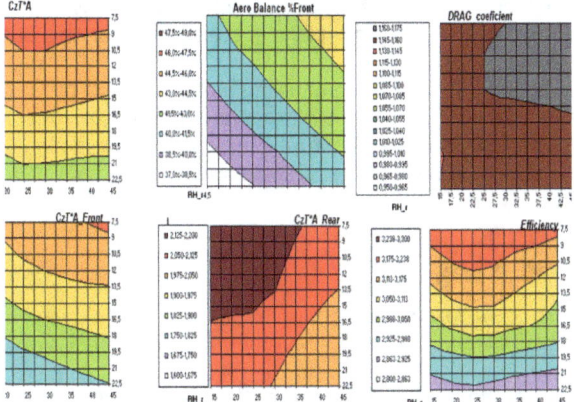

Or we can use a 3D representation:

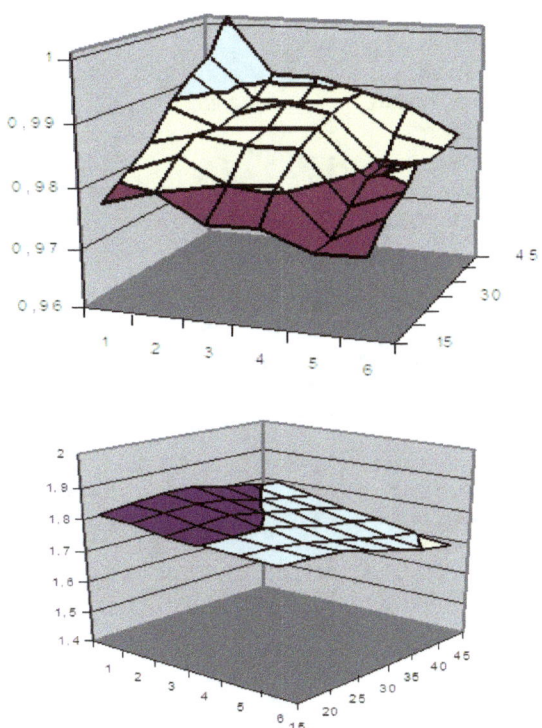

Suppose that the 2 previous 3D graphics represent the front downforce, and each corresponds to a different car:

In the first case, we see that near the intersection of two walls the front downforce is growing rapidly; in the second car, this does not happen.
A glance of the 2 graphs are showing us that the first car is unstable, since it has an area of front and rear heights where the front down force increases abruptly. Anything sharp in relation to aerodynamics, is bad or at least undesirable.

INTERPOLATION

Suppose the following case:
We have 2 dynamic car heights, front and rear: 13 mm and 22 mm.
We have an aero-map of the type table (quantified the downforce in kilos):

	GRADOS	KILOS
FRONT	17º	142,4
REAR	16º	284,2
TOTAL		426,6
COEF TOTAL	2,17	
FRONT	17º	
REAR	17º	
TOTAL		431,2
COEF TOTAL	2,19	

	GRADOS	KILOS
FRONT	17º	141,1
REAR	19º	293,4
TOTAL		434,5
COEF TOTAL	2,21	
FRONT	17º	140,6
REAR	21º	296,2
TOTAL		436,8
COEF TOTAL	2,22	
FRONT	17º	140,2
REAR	23º	297,6
TOTAL		437,8
COEF TOTAL	2,22	
FRONT	17º	140
REAR	25º	297,9
TOTAL		437,9
COEF TOTAL	2,22	

We need to know the parameters of an aerodynamic configuration of 17° front and 21.5° rear.
It is obvious that the table does not provide this information; we interpolate the data; we interpolate by a parable:

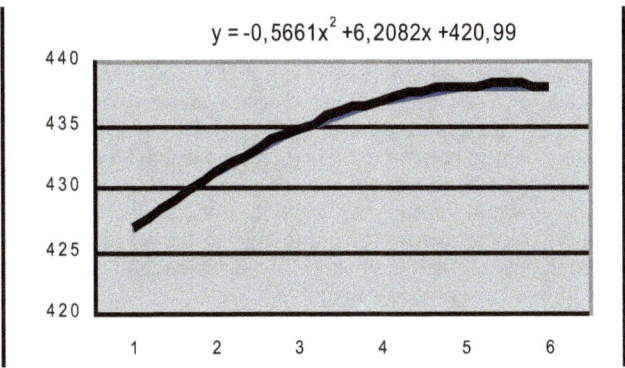

$$y = -0,5661x^2 + 6,2082x + 420,99$$

The parable is a simple but extremely useful function for this type of work; it is not necessary to use polynomials of order 3, 4 or higher; it is sufficient a Grade 2.
Applying the above we get:

	REAR °:	21,5
Total:	437,150	Kg

To compute a certain value, wherever it is, we can turn to extrapolate and interpolate. We can extrapolate and interpolate both horizontally, vertically and diagonally:

CzT

RRH

FRH	5	10	15	20	25	30	35	40	45
5	2.697	2.730	2.80?	2.840	2.?65	2.872	2.?48		
10	2.585	2.652	2.699	2.7??	2.?5?	2.??	2.738	2.724	
15	2.523	2.575	2.619	2.??		2.62?	2.610	2.604	2.602
20		2.501	2.542	2.545	2.541	2.524	2.527	2.505	2.493
25			2.469	2.458	2.459	2.451	2.435	2.422	
30				2.400	2.388	2.381	2.369		

Obviously, the more data we have more reliable inter-extrapolation. We can inter-extrapolate in various directions and obtaining the search data by an average. So, we attenuate the possible error. With few "well distributed" data we can generate all the data in the graph or table. Being well spread is basic because it determines the accuracy of the calculation process of the other values of aero map.

Suppose now that for each angle of incidence front and rear, have a graph of total downforce depending on the front and rear heights; that is, a graph indicating "DF" in which the coordinate axes express front and rear height:

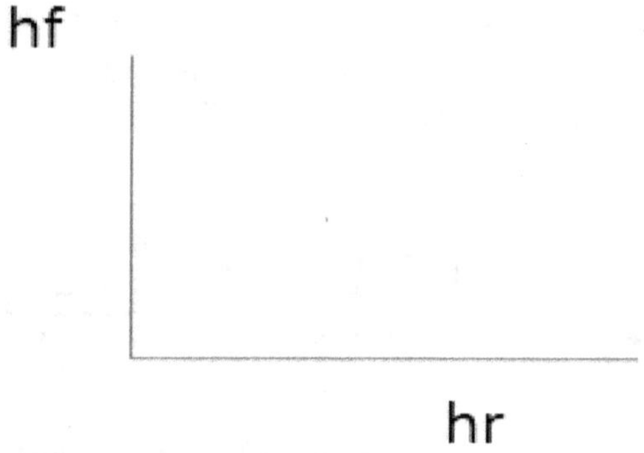

hf

hr

We denote this type of graphics, "G".
If we represent the whole sample space, we get:

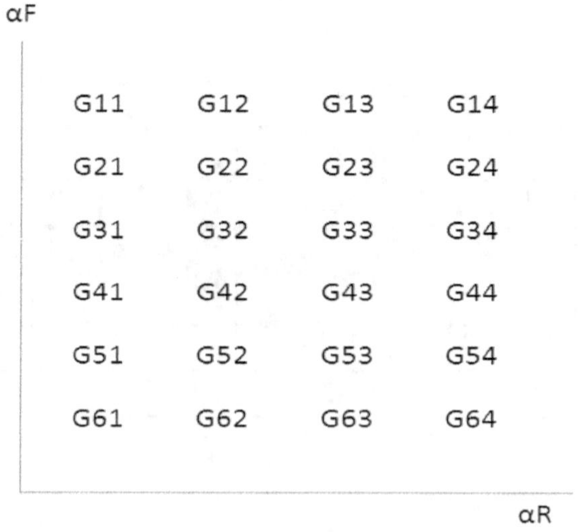

αF

G11	G12	G13	G14
G21	G22	G23	G24
G31	G32	G33	G34
G41	G42	G43	G44
G51	G52	G53	G54
G61	G62	G63	G64

αR

That is, for each front and rear angle of incidence we will have a 2D graphical, formed by points where each point is a 3D graphics.

Now, as happened previously with a graphic as one you must learn to extra-interpolate graphs together. A sort of morphing. This way we will be able to fully ascertain the aero map, based on a known 3D points and these in turn generated by extra-interpolation of a series of points in 2D.

Suppose now these graphs have another component; that is, two angles of incidence and the percentage of opening of a chimney; in this case, we are talking about morphing over 3 dimensions rather than three-dimensional. Note that there are many parameters in a car, not just the incidence angles and aperture of a chimney:

- Front and rear angles of incidence.
- The angle of incidence of the diffuser.
- Percentage of opening of the chimneys.
- Percentage of opening of the pontoons.
- Pressure (shape) of the tires.
- Etc....

Therefore, we are dealing with a problem of morphing in various dimensions.

To solve the three-dimensional problem we can proceed as follows:

Suppose we have certain aero map: it means having a surface for each aerodynamic configuration; our intention to generate other aerodynamic surfaces or data for other aerodynamic configurations, either interpolating or extrapolating; For this we can interpolate or extrapolate values of each aerodynamic configuration; this is:

The line that will link the values of different aerodynamic configurations (pitch) it will also be a parable.

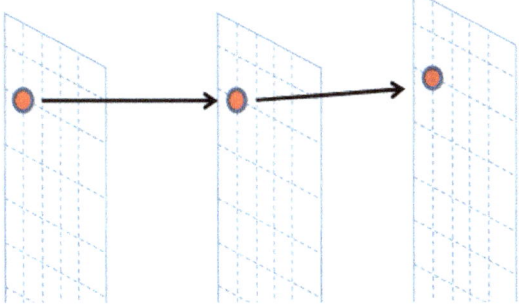

We have already solved the problem of generating absolutely all the aero map of the car, starting with "few" "real" aerodynamic data (on track, through CFD or wind tunnel).

Let us return to the case of a 3D graphics: Suppose we want to calculate a value from others; We can "attach" an area to all known points such that we will have the mathematical equation to calculate the remaining points; such surfaces are called Bezier Surfaces.

Essentially it is a process that approximates a point cloud by a surface called Bezier; the greater the degree of said surface, more accurate and better approximate the points on which it is based:

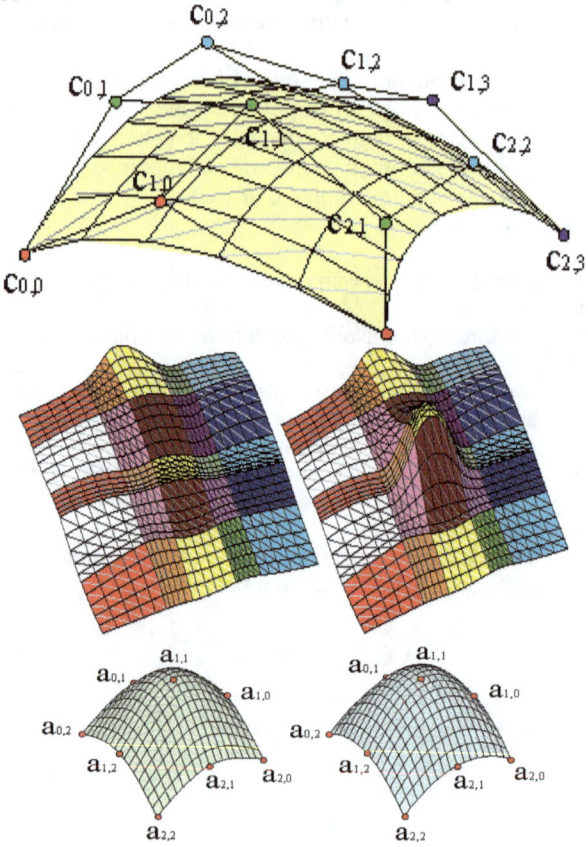

Here we present the code in Mathematica to generate a Bezier surface that is coupled to a point cloud. With the polynomial or generated function we can calculate the value at any point; iterating this process can solve the global problem:
(Points=punts → 3 coordinates correspond to the front and rear heights. The value to study is the total lift coefficient):
punts := {{{5, 5, 2.967}, {5, 10, 2.730}, {5, 15, 2.807}, {5, 20, 2.840}, {5, 25, 2.865}, {5, 30, 2.872}, {5, 35, 2.848}}, etc....

Definition of Bernstein polynomials
B[n_, i_, t_] := Binomial[n, i]*t^i*(1 - t)^(n - i)
Definition Bezier curve
CurvaBezier[poligono_][t_] :=
Sum[BB[Length[poligono] - 1, i, t]*poligono[[i + 1]], {i, 0,
Length[poligono] - 1}] // ExpandAll
Definition of Bernstein polynomials in order to not to give
problems when evaluating them
BB[n_, i_, t_] :=
If[i == 0, If[t == 0, 1, (1 - t)^ n],
If[i == n, If[t == 1, 1, t^n], Binomial[n, i]*t^i*(1 - t)^(n - i)]]
M matrix of values Bernstein polynomials evaluated at given
points in t_i "valort" list
M[n_, k_][valort_] :=
Transpose[Table[BB[n, i, valort[[j + 1]]], {i, 0, n}, {j, 0, k}]]
Matrix:
A[n_, k_][valort_] :=
Inverse[Transpose[M[n, k][valort]].M[n, k][valort]].Transpose[
M[n, k][valort]]
(*n is the degree of Bezier curve (n + 1 control points).*)
(*k is the number of selected points on the curve alpha *)
(* k<= n+1 *)
The "points" list is the vector Q and the result is the vector P
checkpoints curve fitting
approx[puntos_, valort_, n_, k_] :=
A[n, k][valort].Table[puntos[[i + 1]], {i, 0, k}]
punts[[1]]
{{5, 5, 2.967}, {5, 10, 2.73}, {5, 15, 2.807}, {5, 20, 2.84}, {5, 25,
2.865}, {5, 30, 2.872}, {5, 35, 2.848}}
We define the values of t. As the points are equidistant, we put
also values of t equidistant
valort[k_] :=
Table[i/(Length[punts[[k]]] - 1), {i, 0, Length[punts[[k]]] - 1}]
valort[1]
{0, 1/6, 1/3, 1/2, 2/3, 5/6, 1}
We calculate the adjustment control points with a Bezier curve of
degree 5:
approx[punts[[5]], valort[5], 4, Length[punts[[5]]] - 1]
{{25., 15., 2.46895}, {25., 21.25, 2.43605}, {25., 27.5, 2.49481},
{25., 33.75, 2.42475}, {25., 40., 2.42205}}
Now I select 11 points in each curve. So we will have a
rectangular matrix of 6 rows and 11 columns:
Table[Table[
Chop[CurvaBezier[
approx[punts[[j]], valort[j], Length[punts[[j]]] - 1,
Length[punts[[j]]] - 1]][i/10]], {i, 0, 10}], {j, 1, 6}]
puntsnous :=
ListPointPlot3D[punts]

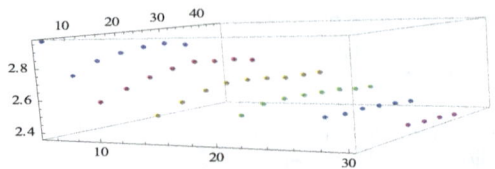

Adjustment surfaces:

B[n_, i_, t_] :=
If[i == 0, If[t == 0, 1, (1 - t)^ n],
If[i == n, If[t == 1, 1, t^n], Binomial[n, i]*t^i*(1 - t)^(n - i)]]
M[n_, k_] := Transpose[Table[B[n, i, j/k], {i, 0, n}, {j, 0, k}]]
A[n_, k_] := Inverse[Transpose[M[n, k]].M[n, k]].Transpose[M[n, k]]
(* n is the order of the Bézier curve (n+1 control points). *)
(* k is the number of points selected in the curve alpha *)
(* k<= n+1 *)
BB[n_, i_, t_] := Binomial[n, i]*t^i*(1 - t)^(n - i)
approx[tabla_, n_, k1_, k2_] :=
A[n, k2].Transpose[
A[n, k1].Table[tabla[[i + 1, j + 1]], {i, 0, k1}, {j, 0, k2}]]

Second attempt with a Bezier surface of degree 5:

Show[ParametricPlot3D[
Evaluate[{1, 1, 10}*
Sum[BB[5, i, u]*BB[5, j, v]*
approx[puntsnous, 5, 5, 10][[j + 1, i + 1]], {i, 0, 5}, {j, 0, 5}]], {u,0, 1}, {v, 0, 1}, Mesh -> None],
ListPointPlot3D[
Table[{1, 1, 10}*punts[[i + 1, j + 1]], {i, 0, 5}, {j, 0, 8}],
Filling -> Bottom]]

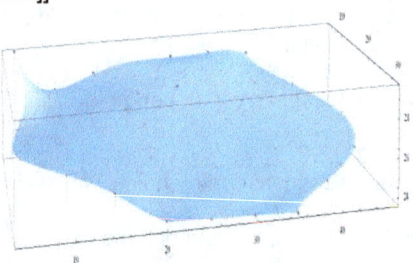

Today we are already able to achieve our CAD models for testing and CFD studies using Bezier surfaces completely; also called Class A surfaces (interpolation point cloud from a 3D scan):

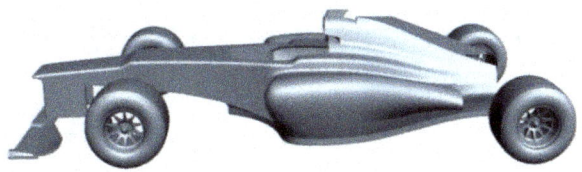

Consider a "real" case: Suppose we have an aero map for a particular aerodynamic configuration and we want to calculate a value for another configuration of heights, which is not in our aero map; we can interpolate-extrapolate values of different ways to obtain the desired result:

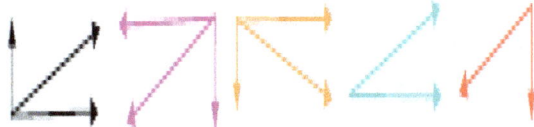

Unknown Value (i,j)=(((2 x Value (i-1,j)- Value (i-2,j))+ (2 x Value (i-1,j+1)-Value (i-2,j+2))+ (2 x Value (i,j+1)- Value (i,j+2)))/3

39,400	39,600	39,900	40,000	40,200
38,300	value(i-2,j)	38,700	value(i-2,j+2)	39,100
0,000	value(i-1,j)	value(i-1,j+1)	37,900	38,100
0,000	unknown	value(i,j+1)	value(i,j+2)	37,100

By doing this we can generate several times the full table of values:

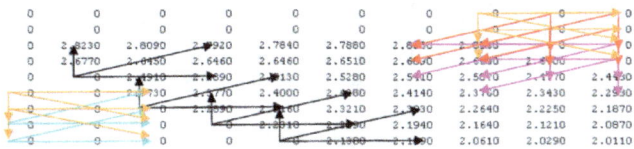

The above calculation scheme is performed, as you can see, with values of the right upper quadrant; You could perform the same operations with the 4 quadrants and take the average of the 4 values obtained. The precision, of course, will be much greater because we draw on more of "known" points

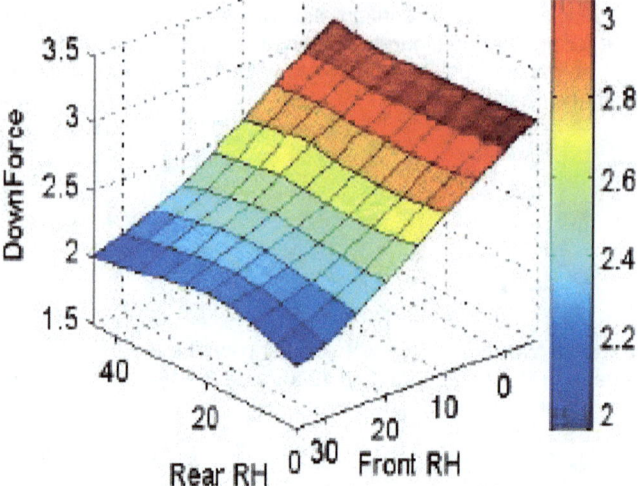

3.1341	3.1121	3.0938	3.0812	3.0735	3.0633	3.0508	3.0529	3.0752	3.1001
2.9990	2.9693	2.9682	2.9095	2.9105	2.9138	2.9187	2.9385	2.9488	2.9440
2.8558	2.8310	2.8090	2.7920	2.7940	2.7880	2.8000	2.8060	2.7960	2.7708
2.7009	2.6770	2.6450	2.6460	2.6460	2.6510	2.6690	2.6600	2.6400	2.5987
2.5343	2.5073	2.4910	2.4890	2.5130	2.5280	2.5410	2.5070	2.4770	2.4480
2.3532	2.3476	2.3730	2.3770	2.4000	2.4280	2.4140	2.3760	2.3430	2.2980
2.1881	2.2204	2.2527	2.2890	2.3160	2.3210	2.3030	2.2640	2.2250	2.1870
2.0509	2.1151	2.1682	2.2127	2.2310	2.2290	2.1940	2.1640	2.1210	2.0870
1.9556	2.0441	2.1127	2.1504	2.1560	2.1380	2.1090	2.0610	2.0290	2.0110

We are able to generate any value we want to know. Set yourself that generated 3D surface is smooth, with no "strangers" spikes that might indicate:

- A bad generated aero map.
- A "bad" behavior of the car.

The availability of "complete" aero map allows us to calculate the following graphs to introduce them in the lap time you're using:

→ We must know the heights of the car at any speed, starting from some initial point (static):

1. Without any acceleration:

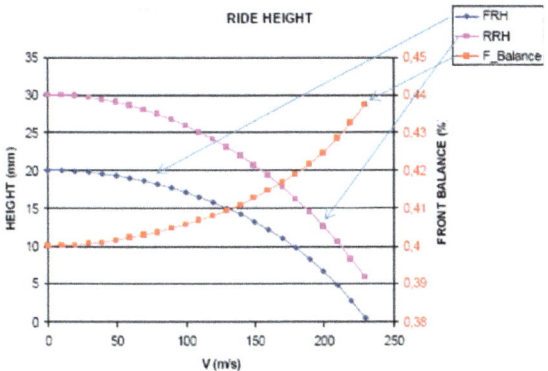

2. In braking.
3. Acceleration.
4. And especially at angles of yaw or lateral acceleration.

For braking and acceleration, we know the weight transfer (front and rear axis and in each wheel), depending on the damping we are using; with these data we will calculate the heights to which is the ground at all times.
This way, we can, at least initially, to define the paths of suspension and the springs to be placed.
Other of the utilities of dispose of the full aero map, is to know a priori, the maximum speed will have our car.

Aerodynamic force: $F_a = \frac{\rho}{2} C_d A_f v^2$

Frictional force: $F_r = (f_0 + f_1 v^2)W$

"f_0", it is the first coefficient of friction, independent of the speed and very small.
"f_1", it is the second coefficient of friction, speed dependent.

$$F_r \sim f_r \cdot W \qquad F_{engine} = F_a + F_r \qquad 1\ HP = 746\ \frac{Nm}{s}$$

$$F_{engine} = \frac{P \cdot 746}{V}$$

"A_f", is front area.
"H_{max}" is engine power.
"P" is the weight.

$$V_{max} = A_1 \left(\left(\sqrt[3]{B_1 + 1} \right) - \left(\sqrt[3]{B_1 - 1} \right) \right)$$

$$A_1 = \sqrt[3]{\frac{H_{max} \cdot \eta_t}{2\left(P \cdot f_r + 0.5 \cdot \rho \cdot C_x \cdot A_f \right)}}$$

$$B_1 = \sqrt[3]{1 + \frac{4 \cdot P^2 \cdot f_r^{\,3}}{27 \cdot H_{max} \cdot \eta_t \cdot \left(f_r + 0.5 \cdot \dfrac{\rho \cdot C_x \cdot A_f}{P}\right)}}$$

$$\eta_{transmission} = \eta_{gearbox} \cdot \eta_{differential}$$

$$F_{wheels} = F_{engine} \cdot \eta_{transmission}$$

ANALYSIS AND CONSEQUENCES

SUDDEN CHANGE

If we look at an aero map and see jumps in values means the car is not stable to movements caused by bumps, bounces or the suspension itself; it can even be dangerous:

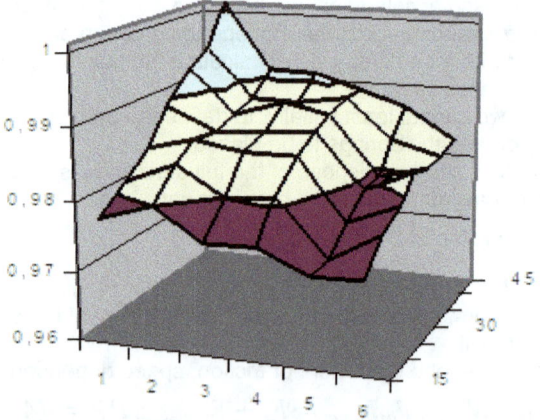

In the above picture we see that around approximately front and rear heights between 2 and 3 mm, and 20 and 30 mm, the measurement expressed by the graph is relatively "uniform".
If we get out of these ranges the measurement increases or decreases faster; This means the car will not be stable to small changes in the heights.
Consider the following case in which we see the opposite: the car is stable to changes in front and rear heights. Variations are uniform and distinguishable:

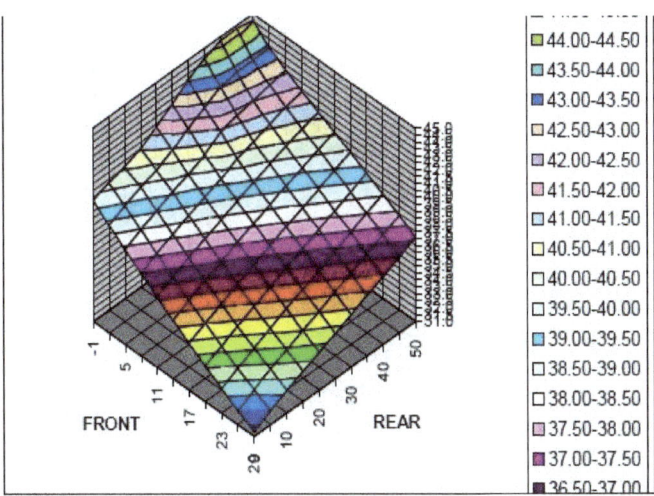

□ 44.00-44.50
□ 43.50-44.00
■ 43.00-43.50
□ 42.50-43.00
□ 42.00-42.50
□ 41.50-42.00
□ 41.00-41.50
□ 40.50-41.00
□ 40.00-40.50
□ 39.50-40.00
□ 39.00-39.50
□ 38.50-39.00
□ 38.00-38.50
□ 37.50-38.00
■ 37.00-37.50
■ 36.50-37.00

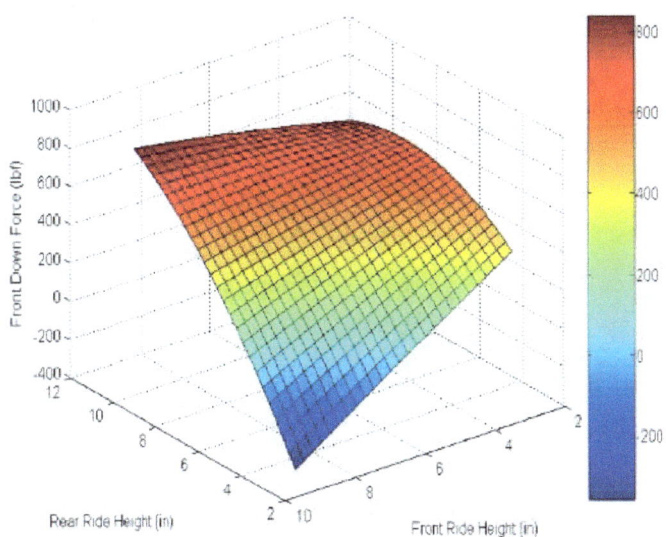

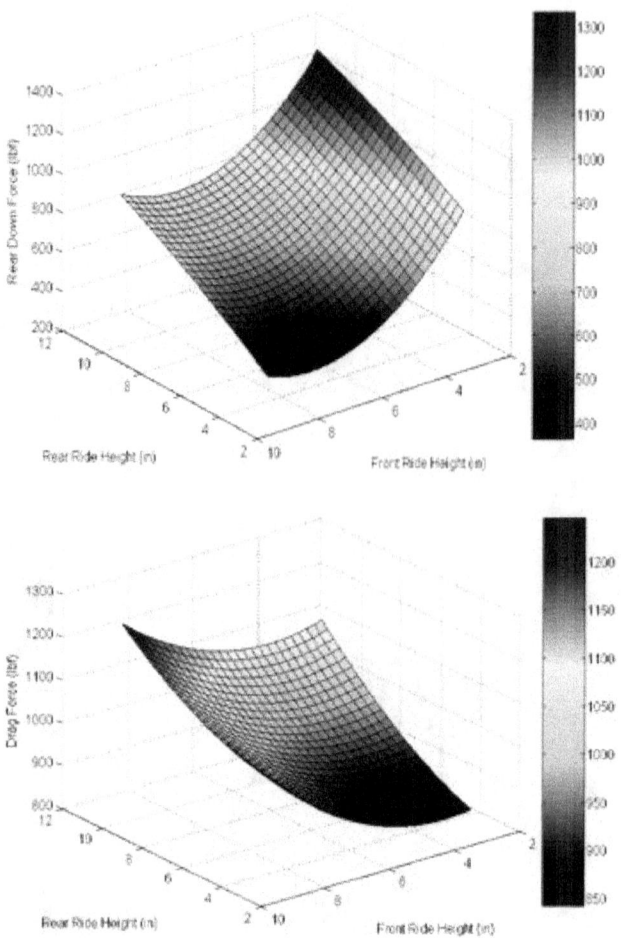

STABILITY AND CALCULATION OF SPRING

Driveability and stability is directly related to a stable behavior within the aeromap; this means that the acceleration phase and in the middle of curve have the same aerodynamic balance. Unstable behavior of the car means big changes in balance under acceleration, straight and curves; hence the "active" suspension that kept the car that always with the same height and angle to the road, is the best because turned the car completely stable aerodynamically speaking (the parallel lines corresponding to positions with the same balance):

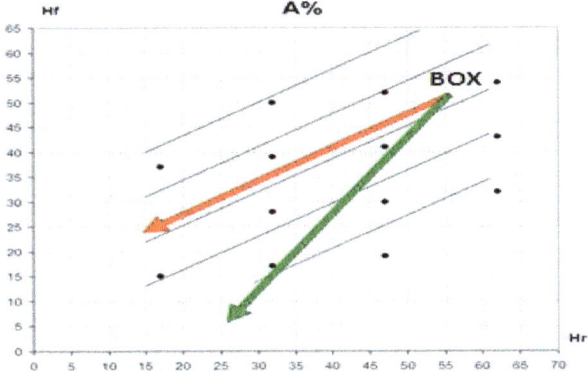

Top line: same iso balance.
Bottom line: unstable – different behavior in balance.
For a specific track, the best setup is one in which all the midpoints of the curves have the same aerodynamic balance; thus does not have "sudden" changes behavior: if it has, we may have "unexpected" under or oversteer under braking, cornering or corner exit:

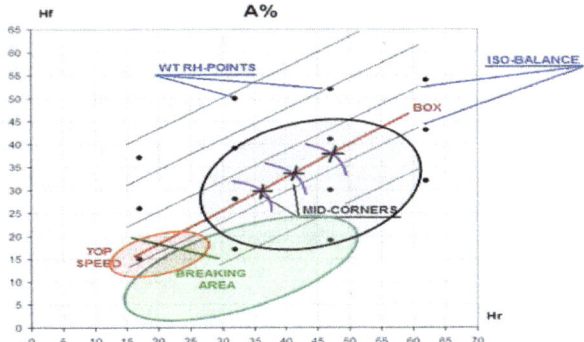

Suppose the best balance for certain car is 45% front; we must find springs so that regardless of the speed, the balance is always 45%; We need each front and rear springs that alter the pitch of the car so that the balance is maintained: the rear height varies more than the front.
For example, consider the following table of balance in terms of the rear (horizontal) and front (vertical) heights:

Aero Balance (%)	5	10	15	20	25	30	35
5	42,7761275	43,2827736	43,6885246	44,0384615	44,472153	45,079013	45,704754
10	41,8984801	42,4731183	42,7883253	43,3588219	43,751771	44,2033124	44,671659
15	41,0635155	41,520979	42,2539313	42,5812116	43,0506478	43,5078379	43,8988095
20		40,3211148	40,8829755	41,4922657	42,1613394	42,7431728	43,2789911
25			39,5069532	40,0633914	40,5861739	41,2233195	42,0611916
30				38,583196	39,2535821	39,8120175	40,3166049

Suppose we want our car has "always" a balance of 40%; heights should be the marked in blue:

Aero Balance (%)	5	10	15	20	25	30	35
5	42,7761275	43,2827736	43,6885246	44,0384615	44,472153	45,079013	45,704754
10	41,8984801	42,4731183	42,7883253	43,3588219	43,751771	44,2033124	44,671659
15	41,0635155	41,520979	42,2539313	42,5812116	43,0506478	43,5078379	43,8988095
20		40,3211148	40,8829755	41,4922657	42,1613394	42,7431728	43,2789911
25			39,5069532	40,0633914	40,5861739	41,2233195	42,0611916
30				38,583196	39,2535821	39,8120175	40,3166049

We know the "law" of variation in front and rear height depending on the speed, so we know what springs we have to use in front and rear.; that is: stiffness front and rear relation; from this relation, we will see that, will be possible to place the natural frequency for each spring. In fact, for calculating each stiffness, the most important is the relation between displacements front and rear.

All this cumbersome and in many cases even complicated procedure (basically for lack of data), is due to changes in downforce of the car, not only in terms of its speed, but also in terms of the inclination (pitch, roll and yaw) . If we could prevent these geometric "variations" the process would be infinitely easier. We will talk about methods to keep the height of the car.

Get an aerodynamic stability at the front axle is relatively easy; but not so on the rear axle; this is due to the complexity of the air flow (and therefore the aerodynamic vibrations) occurring on that area. In the following charts we can see what has been said: we see in the first graph, the line crosses indicating "isobalance desired" approaches very well the lines of the aero map; while in the second graph (rear downforce), not so:

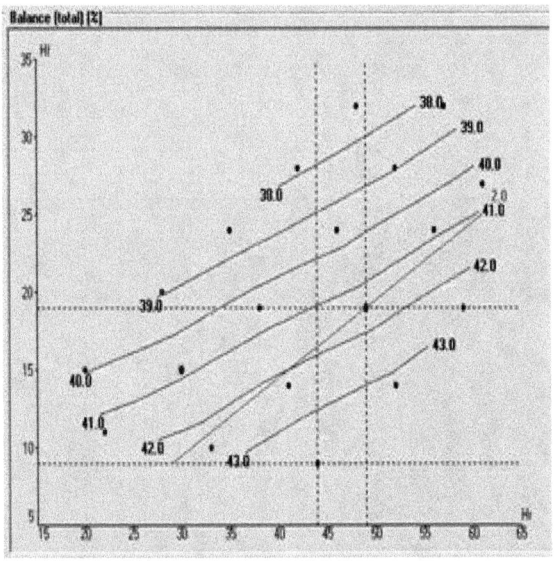

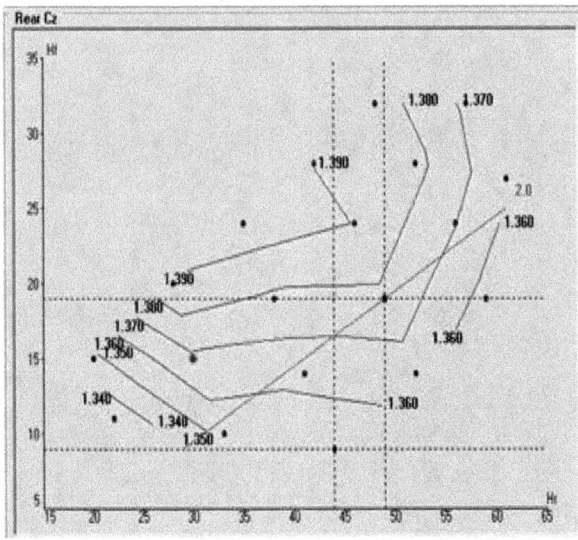

This fact can generate oversteer or understeer.
As an example in order to calculate the two springs, we can work as that:

STARTING FOR THE FOLLOWING ANALYSIS: PAG 1

WE HAVE

$$① \quad f_1^2 \cdot \delta_1 = f_2^2 \cdot \delta_2$$

THIS IS THE RELATIONSHIP BETWEEN
THE FREQUENCY AND SUSP. DEFLECTION
IN THE CASE OF A FORMULA CAR
WITH HIGH DOWNFORCE AS F1/F3000
WE HAVE

$$② \quad \delta_2 = 2 \, \delta_1$$

THEN FROM ①

$$f_2 = f_1 \cdot \sqrt{\frac{\delta_1}{\delta_2}}$$

$$= f_1 \sqrt{\frac{\delta_1}{2\delta_1}} = f_1 \cdot \sqrt{\frac{1}{2}}$$

$$\boxed{f_2 = (.707) \cdot f_1}$$

— HAVING TYRES WITH FREQUENCY OF

FRONT TYRE — $f_{TF} = 5 \ \{Hz\}$

REAR TYRE — $f_{TR} = 5.5 \ \{Hz\}$

— THEN TO BE SURE THAT WE WILL HAVE THE
SUSP WORKING AT LOWER FREQUENCY THAT THE
TYRE NATURAL FREQUENCY WE FIX:

$$\boxed{f_1 = 4.5 \ \{Hz\}}$$

$$\Rightarrow \quad f_2 = \ \text{~~HIGH~~} \ (.707) \cdot f_1 = (.707) \cdot (4.5)_0$$

$$\boxed{f_2 = 3.18 \ \{Hz\}}$$

WHERE:

f_1 = FRONT SUSP FREQUENCY

f_2 = REAR SUSP FREQUENCY

δ_1 = FRONT SUSP DEFLECTION

δ_2 = REAR SUSP DEFLECTION

WORKING WITH A SIMPLIFIED ANALYSIS, WE HAVE:

SPRING RATE FOR EACH SPRING.

- FRONT SPRING:

$$K_{R_1} = \frac{(.002) \cdot f^2 \cdot w \cdot b \cdot X_1^2}{\ell} \quad \left\{ \frac{kg}{mm} \right\}$$

- REAR SPRING:

$$K_{R_2} = \frac{(.002) \cdot f^2 \cdot w \cdot a \cdot X_2^2}{\ell} \quad \left\{ \frac{kg}{mm} \right\}$$

WHERE:

f = FREQUENCY IN (Hz)

w = SPRUNG MASS (Kg)

b = DISTANCE TO G.C. TO REAR AXLE

a = "   "   "   " FRONT AXLE

ℓ = WHEELBASE

K_{R_1} = SPRING RATE FRONT SPRING

K_{R_2} = "   " REAR SPRING

X_1 = INSTALATION RATIO FRONT SUH.

X_2 = "   " REAR SUSP.

FOR THE F 3000:

#) ~~PARAMETERS~~:

$X_1 = 1.20$; $X_2 = 1.45$ $\qquad$ $f_1 = 4.5\ (Hz)$

$l \approx 2700\ (mm)$ $\qquad\qquad\qquad\quad$ $f_2 = 3.18\ (Hz)$

$a = 1620\ (mm)$

$b = 1080\ (mm)$

$P \approx 500 + 100 + 75 = 675\ (kg)$

$S_w = 110\ (kg)$ UNSPRUNG MASS

$\Rightarrow W = 675 - 110 = 565\ (kg)$

THEN:

a) FRONT SUSP:

$$K_{R_1} = \frac{(.002) \cdot f_1^2 \cdot W \cdot b \cdot X_1^2}{l} = \frac{(.002) \cdot (4.5)^2 \cdot 565 \cdot (1080) \cdot (1.2)^2}{2700}$$

$\boxed{K_{R_1}} = 13.18\ (kg/mm) = \boxed{737\ \left(\frac{lbf}{in}\right)}$

TO START WE FIX:

$$K_{R_1} = 800\ \left(\frac{lbf}{in}\right) = 14.3\ \left(\frac{kg}{mm}\right)$$

b) REAR SUSP:

$$K_{R_2} = \frac{(.002) \cdot f_2^2 \cdot W \cdot a \cdot X_2^2}{l} = \frac{(.002) \cdot (3.18)^2 \cdot 565 \cdot (1620) \cdot (1.45)^2}{2700} =$$

$\boxed{K_{R_2}} = 14.41\ (kg/mm) = 806\ \left(\frac{lbf}{in}\right)$

To calculate the springs perfectly, we must take into account the deflection of the tire; It is a very complicated issue but we can make an initial simplification: knowing the constant "Kw" of the tire and constant "Ks" of the suspension; We suppose that are parallel; We can then calculate the global constant:

$$\frac{1}{Ktotal} = \frac{1}{Kw} + \frac{1}{Ks}$$

Having an asymmetrical flow may present problems of instability, as the yaw moment "Mz" tend to rotate the vehicle about the vertical axis "Z". If this increase tend to reduce the angle of incidence, the vehicle is aerodynamically stable; the contrary causes aerodynamic instability; that is to say:

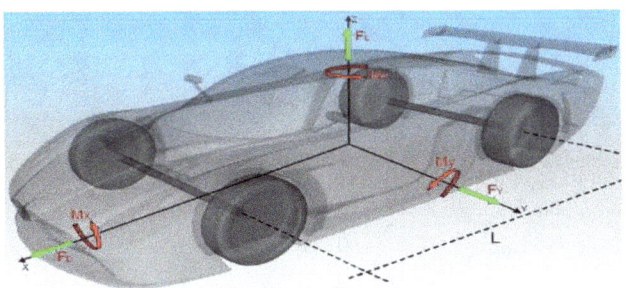

$$\text{Aerodynamically stable vehicle} = \frac{\Delta C_{mz}}{\Delta Incidence} < 0$$

An aerodynamically stable vehicle is one that, for a small perturbation makes our car has a slight inclination, either side wind or other circumstances the vehicle creates a counter point to the wind, which counteracts this wind bringing it to its original point. The more stable our vehicle, the faster it will return to its place of origin.

To see if a vehicle is aerodynamically stable to lateral winds, we have to see if the increase in the yaw moment according to the wind angle that is disturbing it, is smaller than zero. On the contrary our car would be aerodynamically unstable to lateral wind.

Is possible to know if are aerodynamically stable, from their inertia moments and incidence angles:

Very unstable:

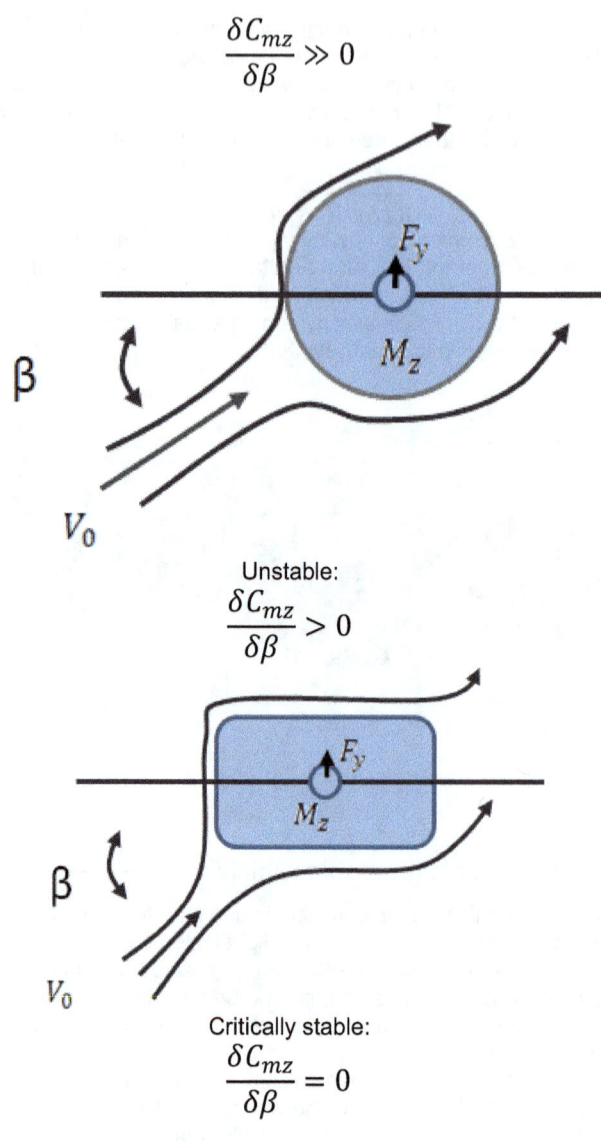

$$\frac{\delta C_{mz}}{\delta \beta} \gg 0$$

Unstable:
$$\frac{\delta C_{mz}}{\delta \beta} > 0$$

Critically stable:
$$\frac{\delta C_{mz}}{\delta \beta} = 0$$

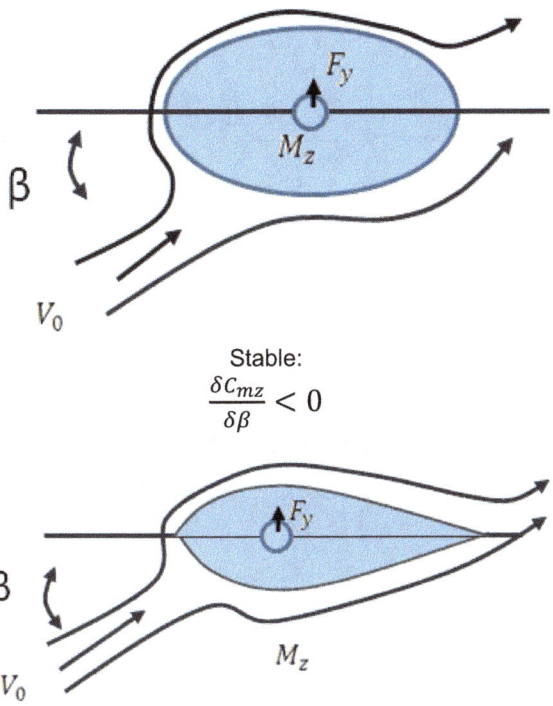

Stable:

$$\frac{\delta C_{mz}}{\delta \beta} < 0$$

The above formulas always be met when the forces we obtain are on the application point of the center of gravity. In our simulation the forces that have been obtained have been referenced on the front axle, so the above formulas would not be entirely true to not refer to the same point. It then chooses to see the point where the vehicle is critically stable, being unstable if you go from this point.

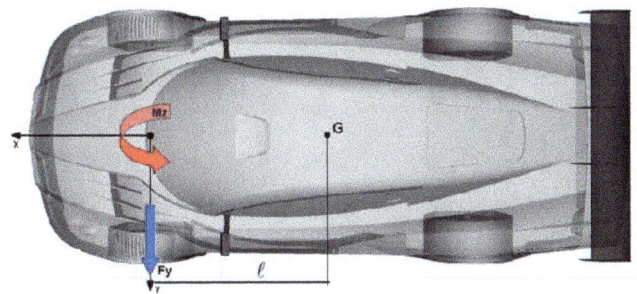

$$M_G = M_Z + F_y \cdot l$$

$$\frac{\Delta M_G}{\Delta \alpha} = 0 \quad \longrightarrow \quad \frac{\Delta M_Z + \Delta F_y \cdot l}{\Delta \alpha} = 0 \quad \longrightarrow \quad l = \frac{-\Delta M_Z}{\Delta F_y}$$

OTHER CONSIDERATIONS AND ANALYSIS

Let us take three circuits:

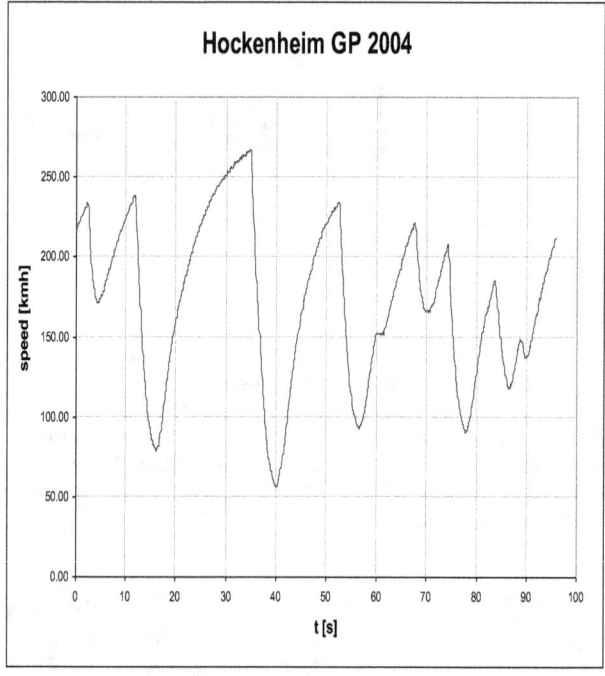

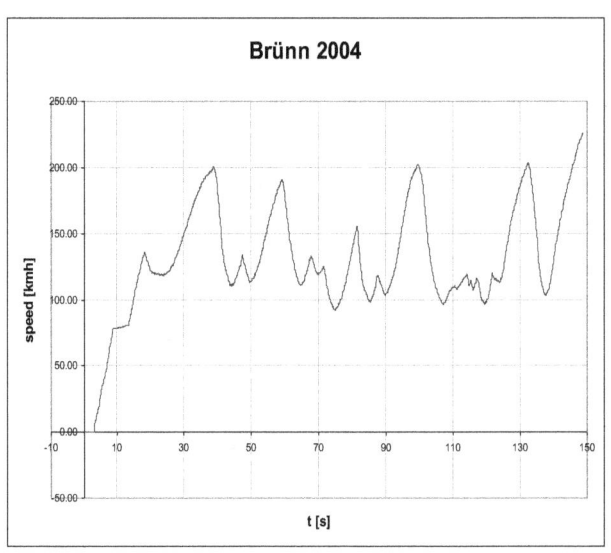

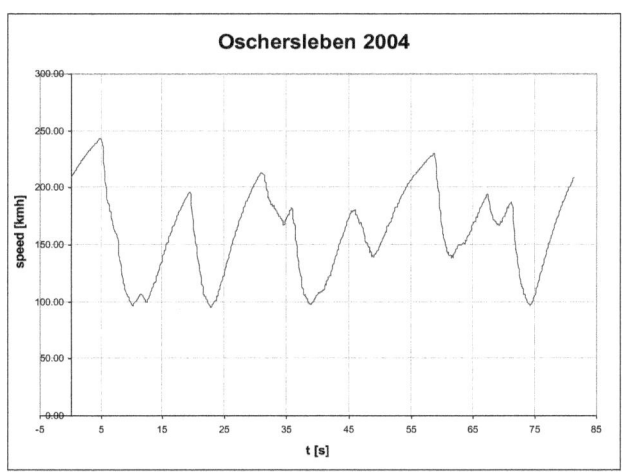

In these three circuits, we see the front and rear heights reached during each lap:

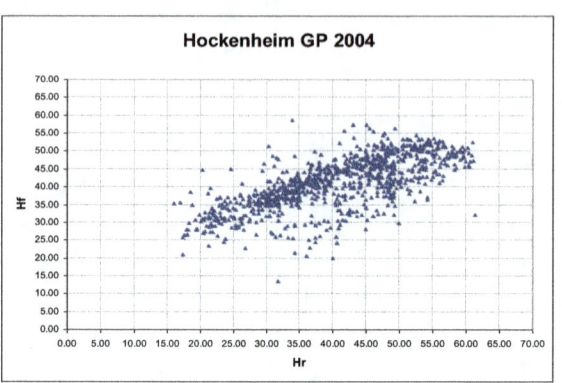

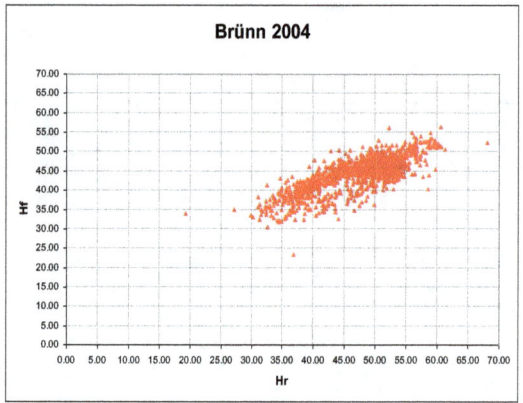

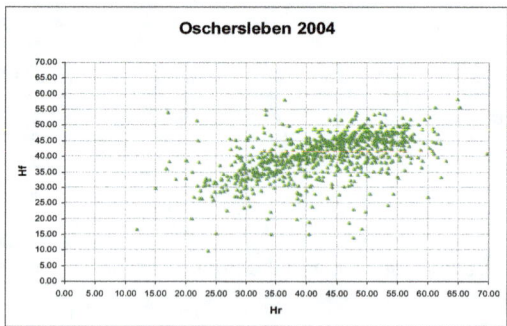

The dispersion of data for these graphs may indicate us many things such as that it is a heavy braking circuit, it is when there are more variations of front and rear heights or displacement of the shock absorbers; the second circuit is "uniform". But all of them are close to a straight line (aerodynamic stability).
First, we must identify the curves:

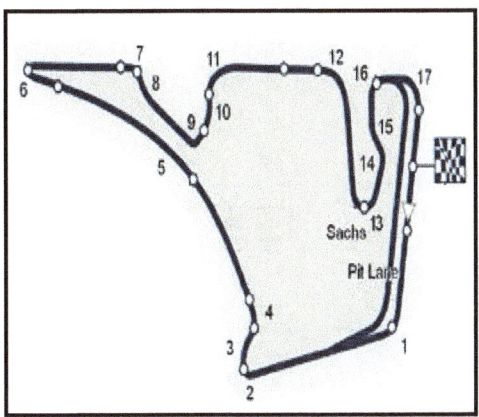

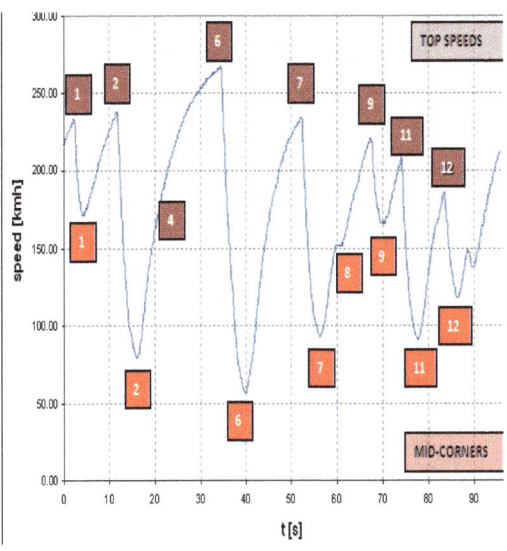

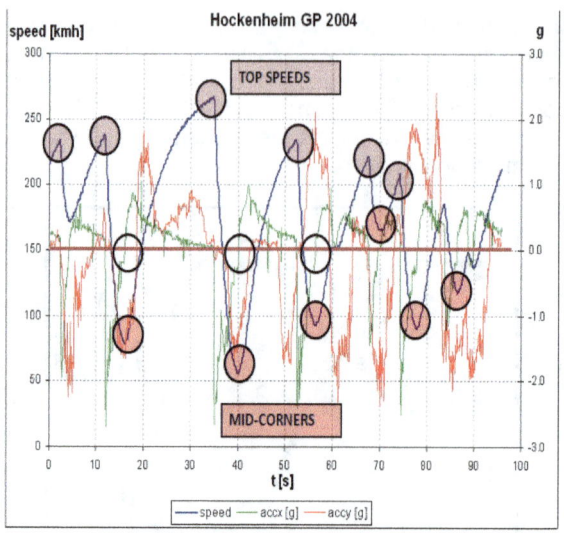

Now we represent the heights at these points in a straight, middle curve and braking:

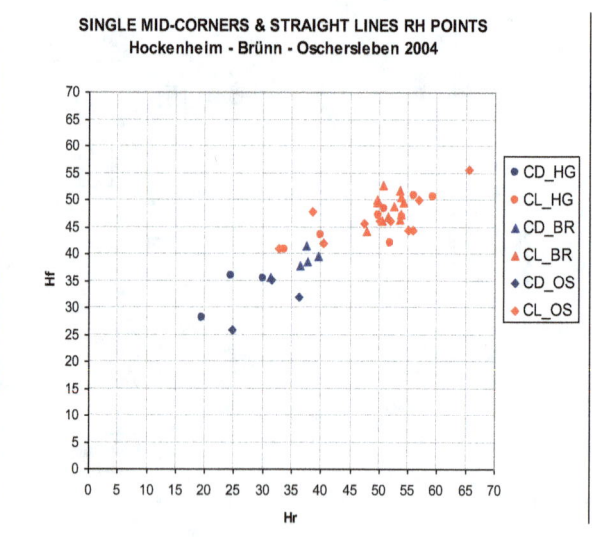

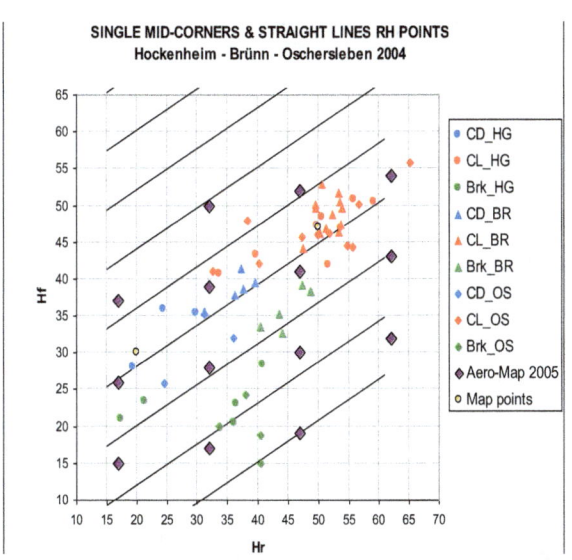

SINGLE MID-CORNERS & STRAIGHT LINES RH POINTS
Hockenheim - Brünn - Oschersleben 2004

Legend:
- CD_HG
- CL_HG
- Brk_HG
- CD_BR
- CL_BR
- Brk_BR
- CD_OS
- CL_OS
- Brk_OS
- Aero-Map 2005
- Map points

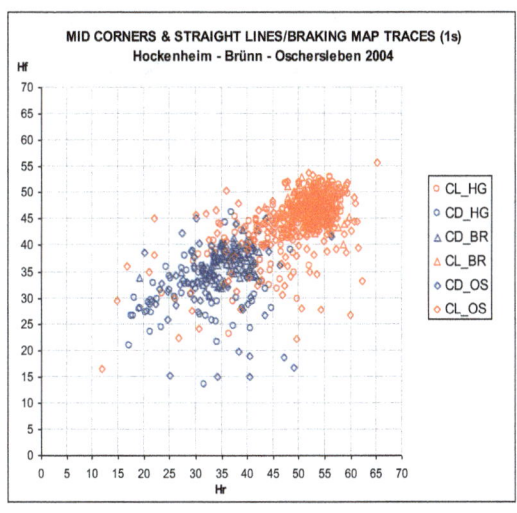

MID CORNERS & STRAIGHT LINES/BRAKING MAP TRACES (1s)
Hockenheim - Brünn - Oschersleben 2004

Legend:
- CL_HG
- CD_HG
- CD_BR
- CL_BR
- CD_OS
- CL_OS

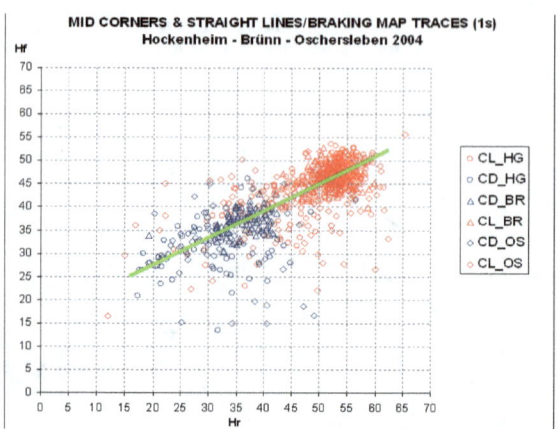

Therefore, the car, all circuits studied is stable aerodynamically speaking.

Here's a typical case: the car has several front wings or front positions to achieve and reach the whole rank of "dynamic" configurations; it is not common, but can occur in certain categories: Every time we change the wing or configuration to achieve some aerodynamic values, we can cause a jump; or what is the same: a jump or "empty" of aerodynamic values; in this case: 3 spoilers:

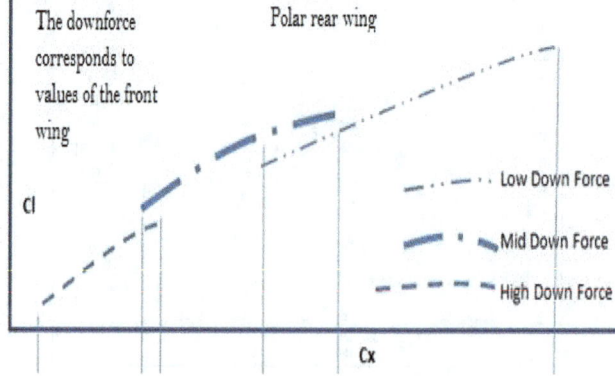

We note that there are jumps or rather, there are overlapping areas; it is necessary to consider these overlaps. Consider the downforce of the front wing:

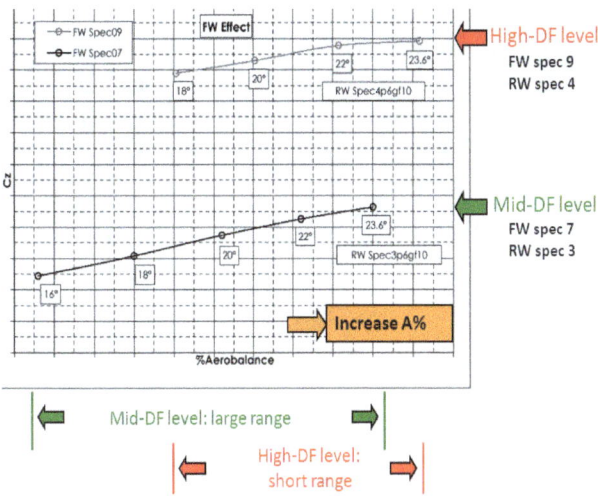

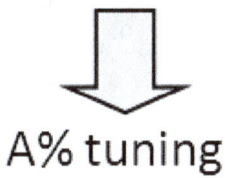

Different Aero-balance range in function of the DF level (RW)

⬇

A% tuning

AERODYNAMIC REBALANCING

On countless occasions, since we do not have nearly ever the complete aero map, but only a series of values of a number of configurations, we want to know:

- Variations of the "CP" on function of variations of the aerodynamic configuration.
- Variations of the drag and downforce, depending on variations in aerodynamic configuration.

To do this, we can make a linear interpolation:
The law of re-balancing calculates the correct aerodynamic coefficients comparing them with the same objective:

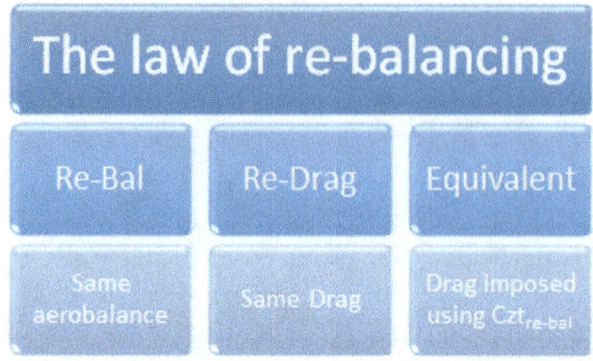

Re-Bal	Re-Drag	Equivalent
Same aerobalance	Same Drag	Drag imposed using Czt$_{re-bal}$

- In type passenger cars and sports cars in general, to modify the "CP" and the downforce rear wing is modified.
- In Formula 1 car type, modification of the front wing is used to modify the "CP" and rear spoiler for the downforce and drag.

Laws of Re-Balancing:
Front Wing Law:

$$\alpha = \frac{\partial C_z}{\partial A\%} ; \beta = \frac{\partial C_x}{\partial A\%}$$

Rear Wing Law:

$$\gamma = \frac{\partial C_z}{\partial C_x} ; \delta = \frac{\partial A\%}{\partial C_x}$$

For example, to see this interpolation between 2 points:

$$\frac{\partial C_{zt}}{\partial A\%} = \frac{C_{z\,FLAP\,21°} - C_{z\,FLAP\,19°}}{Bal_{FLAP\,21°} - Bal_{FLAP\,19°}}$$

$$\frac{\partial C_{xt}}{\partial A\%} = \frac{C_{x\,FLAP\,21°} - C_{x\,FLAP\,19°}}{Bal_{FLAP\,21°} - Bal_{FLAP\,19°}}$$

The laws of re-balancing are calculated with a simple linear interpolation with at least 3 measuring points in wind tunnel (near the target to be developed).

Re-drag equations: C_z measured, C_x measured and A% measured, are weighted values (measured in wind tunnel):

$$C_{zt\,\text{RE-DRAG}} = C_{zt\,\text{MEASURED}} + \frac{\partial C_{zt}}{\partial C_x}\left(C_{x\,\text{TARGET}} - C_{x\,\text{MEASURED}}\right)$$

$$A\%_{\text{RE-DRAG}} = A\%_{\text{MEASURED}} + \frac{\partial A\%}{\partial C_x}\left(C_{x\,\text{TARGET}} - C_{x\,\text{MEASURED}}\right)$$

$$Eff_{\text{RE-DRAG}} = \frac{C_{zt\,\text{RE-DRAG}}}{C_{xt\,\text{TARGET}}}$$

Re-Balance: C_x y %A targets:

$$C_z equiv = C_z re - drag + \alpha(\%A - t\arg et - \%A - re - drag)$$
$$C_x t\arg et = C_x re - drag + \beta(\%A - t\arg et - \%A - re - drag)$$

Equivalent downforce:

$$C_z equiv = C_z measured + \frac{\alpha - \beta\gamma}{1 - \beta\delta}(\%A - t\arg et - \%A - measured) +$$
$$+ \frac{\gamma - \alpha\delta}{1 - \beta\delta}(C_x - t\arg et - C_x measured)$$

$C_{x\,TARGET}$ imposed

$$C_{zt\,\text{RE-DRAG}} = C_{zt\,\text{RE-BAL}} + \left[\frac{\partial C_{zt}}{\partial C_x} - \left(\frac{\partial C_{zt}}{\partial A\%} \cdot \frac{\partial A\%}{\partial C_x}\right) + \frac{\partial C_{zt}}{\partial C_x} \cdot \frac{\partial A\%}{\partial C_x} \cdot \frac{\partial C_x}{\partial A\%}\right] \cdot \left(C_{x\,\text{TARGET}} - C_{x\,\text{RE-BAL}}\right)$$

$$Eff_{\text{EQUIVALENT}} = \frac{C_{zt\,\text{EQUIVALENT}}}{C_{zt\,\text{TARGET}}}$$

									LH	Rh	Re Balance	Equivalent	
Effc	Bal	Czt	Czf	Czr	Czfw	Czrw	Czb	Czt	Dep1	Dep2	CzBal	Czeq	EffEq
3,794	48,46	2,138	1,036	1,102	0,029	0,088	0,448	0,565	0,283	0,273	2,138	2,138	3,784
3,789	48,48	2,133	1,034	1,099	0,03	0,087	0,446	0,563	0,282	0,274	2,132	2,149	3,803
3,762	47,92	2,118	1,015	1,103	0,028	0,087	0,446	0,563	0,281	0,274	2,135	2,155	3,814
3,716	47,33	2,096	0,992	1,104	0,028	0,087	0,449	0,564	0,281	0,276	2,131	2,147	3,799
3,797	49,76	2,092	1,041	1,051	0,029	0,086	0,436	0,551	0,279	0,272	2,052	2,159	3,820
3,788	51,57	2,034	1,049	0,985	0,029	0,085	0,423	0,537	0,275	0,269	1,937	2,148	3,801
3,792	48,57	2,135	1,037	1,099	0,029	0,087	0,448	0,563	0,282	0,274	2,132	2,147	3,801
3,812	46,90	2,211	1,037	1,174	0,029	0,092	0,459	0,58	0,286	0,277	2,259	2,146	3,798

α	0,031		γ	3,891
β	-0		δ	-115,84
$\alpha 1$	0,031		$\gamma 1$	3,891
$\beta 1$	-0,0008		$\delta 1$	-115,84

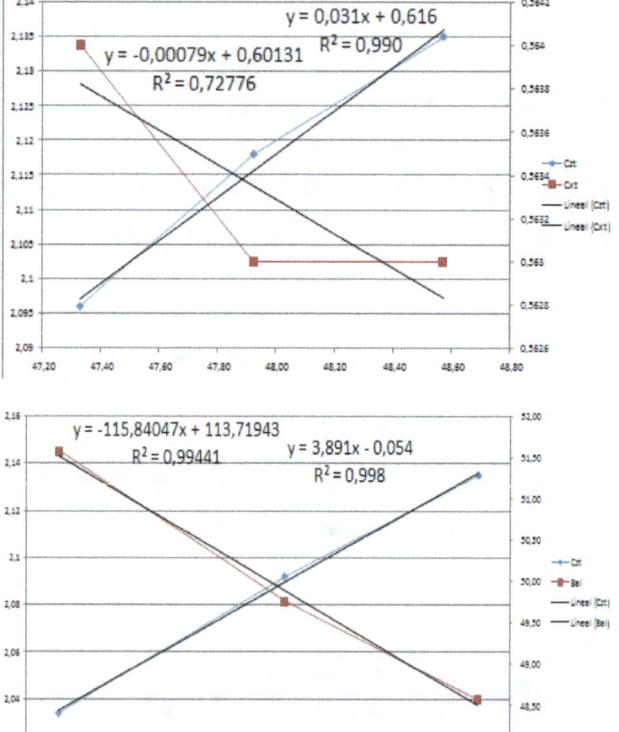

DATA TO BE PROVIDED BY THE AERO-MAP

The aero map must provide all aerodynamic values, corresponding to all configurations of the car; in essence, that is.

- Task: providing a significant amount of data which will be properly re-arranged in a summary table displaying the available car set-ups and used in an interpolation tool aimed at set-up customization.
- Procedure (Sports car example): *PARAMETRICAL SURVEY:*
  - o Identification of down force levels based on the most performing front down force add-ons (fences/strakes, flicks, louvers, guide vanes, etc.).
  - o Mapping of the rebalancing devices parameter-wise (rear wing: main/flap) aimed at achieving the most efficient and uniform rear down force distribution.
  - o Mapping of the cooling devices (louvers, extractors) aimed at identifying the cooling levels for each down force level.
  - o Building and testing the ultimate rebalanced set-ups matching down force and cooling requirements.

The number of involved parameters is variable and mutual correlated effect have to be accounted for!

- Further parameters: the car set-up could be intended as a description of the dynamic behaviour of the car, not just as a combination of the configuration parameters only
- Assuming the previous standard aeromapping survey data these are some attitude parameters:
  - o Yaw.
  - o Steering.
  - o Roll.
- These tests can be combined to the standard aero map (Heave & Pitch) in order to provide full descriptions (steady state) used as input for the performance (laptime) simulation.

N.B. The set of variable range quantities has to be selected within the most representative attitudes of the racing car performance (fast & slow corners, end of straight line).
  → Important notation:

We have already commented but we insist: the aero map must contain details of the car in all possible positions and actions; in the section Post Rig we will see the importance of knowing these values "dynamic" or "transiently"; aerodynamic data of the car at certain angles of yaw, is absolutely necessary and we must invest much time in their calculation.

9.9. EXAMPLES OF INTERPOLATION EXTRAPOLATION: STUDIES STRATEGIES FOR OBTAINING A FULL AERO MAP

Example: representing the coefficient of total downforce, for a certain aerodynamic configuration of a car:

Cz	6	8	10	12	14	16	18	20	22	24	26	28	30
5,0	1,603	1,664	1,702	1,765	1,828	1,850	1,873	1,895	1,917	1,924	1,932	1,939	1,946
7,5	1,665	1,702	1,737	1,769	1,800	1,818	1,835	1,853	1,871	1,873	1,875	1,877	1,880
10,0	1,793	1,778	1,772	1,772	1,772	1,785	1,798	1,811	1,824	1,821	1,819	1,816	1,813
12,5	1,798	1,773	1,766	1,760	1,755	1,765	1,775	1,785	1,796	1,795	1,794	1,793	1,793
15,0	1,806	1,774	1,760	1,749	1,738	1,745	1,752	1,760	1,767	1,768	1,770	1,771	1,772

When working with a race car, we barely have aerodynamic data; in most cases, a lack of data is complete. For this reason, the goals we set are:
- Develop strategies to calculate points from other acquaintances. In this way, among other things, we define strategies for optimal use of test time in the wind tunnel.
- Calculate the full aero map.

Let us work the aero map (total downforce coefficient) for the Dallara F-302, for example; it would be to see if we get the same results or strategies with other car models.
See the working zone (in white); the other measures of heights, do not use:

Cz	6	8	10	12	14	16	18	20	22	24	26	28	30
5	1,603	1,664	1,702	1,765	1,828	1,85	1,873	1,895	1,917	1,924	1,932	1,939	1,946
7,5	1,665	1,702	1,737	1,769	1,8	1,818	1,835	1,853	1,871	1,873	1,875	1,877	1,88
10	1,793	1,778	1,772	1,772	1,772	1,785	1,798	1,811	1,824	1,821	1,819	1,816	1,813
12,5	1,798	1,773	1,766	1,76	1,755	1,765	1,775	1,785	1,796	1,795	1,794	1,793	1,793
15	1,806	1,774	1,76	1,749	1,738	1,745	1,752	1,76	1,767	1,768	1,77	1,771	1,772

This work area is fairly flat; Let us draw the columns from the rear 14 mm height:

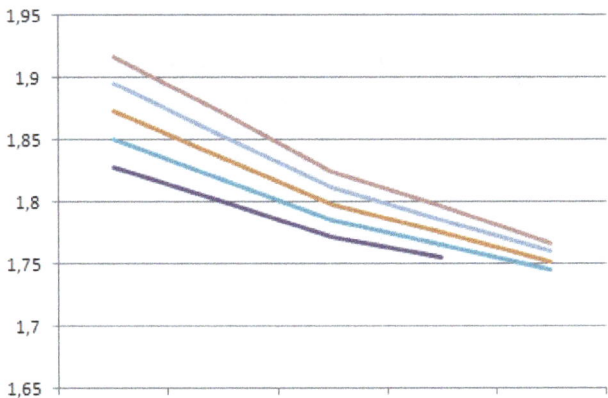

Let us represent the rows now:

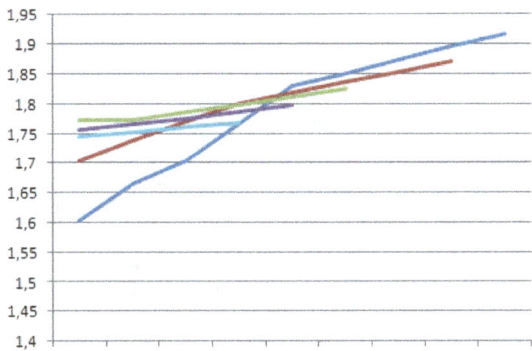

We can observe certain "linearity".
We need disclose a method by which we are able to calculate all values with the least possible error.
We will study the mistake we can make using different methods and strategies. Our work table is:

1,603	1,664	1,702	1,765	1,828	1,85	1,873	1,895	1,917
	1,702	1,737	1,769	1,8	1,818	1,835	1,853	1,871
			1,772	1,772	1,785	1,798	1,811	1,824
				1,755	1,765	1,775	1,785	1,796
					1,745	1,752	1,76	1,767

First, assume that the table identified by a flat.
Flat ax+by+cz+d=0 passing through 3 points (p1(1), p1(2), p1(3)) , (p2(1), p2(2), p2(3)) , (p3(1), p3(2), p3(3)):

$$a = (p2(2) - p1(2)) * (p3(3) - p1(3)) - (p2(3) - p1(3)) * (p3(2) - p1(2))$$

$$b = (p2(3) - p1(3)) * (p3(1) - p1(1)) - (p2(1) - p1(1)) * (p3(3) - p1(3))$$
$$c = (p2(1) - p1(1)) * (p3(2) - p1(2)) - (p2(2) - p1(2)) * (p3(1) - p1(1))$$
$$d = - p2(1) * a - p2(2) * b - p2(3) * c$$

Or what is the same: (x1,y1,z1) , (x2, y2, z2) , (x3, y3, z3):

$$\begin{vmatrix} x - x_1 & y - y_1 & z - z_1 \\ x_2 - x_1 & y_2 - y_1 & z_2 - z_1 \\ x_3 - x_1 & y_3 - y_1 & z_3 - z_1 \end{vmatrix} = 0$$

Line through 2 points (x1,f(x1)) , (x2,f(x2)) → f(x) = ax+b

$$a = \frac{f(x2) - f(x1)}{x2 - x1}$$

$$b = f(x1) - x1 \frac{f(x2) - f(x1)}{x2 - x1}$$

First, we will see the error committed by "assuming" that the table is flat; we know how to interpolate and extrapolate values to create "whole" aero map.

To suppose that fits neatly into a flat, it is the same as assuming that each vertical and horizontal line are straight. Suppose we want to calculate the value marked in red:

We can calculate this by interpolating and extrapolating other values of the table value.

- Scenario 1: given "normally" the scarcity of values to work comfortably, we use polynomials of grade 1 and grade 2 at most. In the case of having sufficient data, logically, higher degree polynomials always be exact.
- Scenario 2: If interpolate, the error is always less.

Interpolation of grade 1 from the points with green background:

Cz	6	8	10	12	14	16	18	20	22	24	26	28	30
5	1,603	1,664	1,702	1,765	1,828	1,85	1,873	1,895	1,917	1,924	1,932	1,939	1,946
7,5	1,665	1,702	1,737	1,769	1,8	1,818	1,835	1,853	1,871	1,873	1,875	1,877	1,88
10	1,793	1,778	1,772	1,772	1,772	1,785	1,798	1,811	1,824	1,821	1,819	1,816	1,813
12,5	1,798	1,773	1,766	1,76	1,755	1,765	1,775	1,785	1,796	1,795	1,794	1,793	1,793
15	1,806	1,774	1,76	1,749	1,738	1,745	1,752	1,76	1,767	1,768	1,77	1,771	1,772

We calculate the line through the "green" points and interpolate the "red" value:

a)

Cz	6	8	10	12	14	16	18
5	1,603	1,664	1,702	1,765	1,828	1,85	1,873
7,5	1,665	1,702	1,737	1,769	1,8	1,818	1,835
10	1,793	1,778	1,772	1,772	1,772	1,785	1,798
12,5	1,798	1,773	1,766	1,76	1,755	1,765	1,775
15	1,806	1,774	1,76	1,749	1,738	1,745	1,752

x1=7.5   x2=12.5   f(x1)=1.835   f(x2)=1.775
Calculated value: f(10)= 1.805   Real value= 1.798
Error= 0.4%

b)

14	16	18	20	22
1,828	1,85	1,873	1,895	1,917
1,8	1,818	1,835	1,853	1,871
1,772	1,785	1,798	1,811	1,824
1,755	1,765	1,775	1,785	1,796
1,738	1,745	1,752	1,76	1,767

x1=16   x2=20   f(x1)=1.785   f(x2)=1.811
Calculated value: f(10)= 1.798   Real value= 1.798
Error= 0%

c)

16	18	20
1,85	1,873	1,895
1,818	1,835	1,853
1,785	1,798	1,811
1,765	1,775	1,785
1,745	1,752	1,76

x1=16   x2=20   f(x1)=1.765   f(x2)=1.853
Calculated value: f(10)= 1.809   Real value= 1.798
Error= 0.6%

d)

16	18	20
1,85	1,873	1,895
1,818	1,835	1,853
1,785	1,798	1,811
1,765	1,775	1,785
1,745	1,752	1,76

x1=16   x2=20   f(x1)=1.818   f(x2)=1.785
Calculated value: f(10)= 1.8015   Real value= 1.798
Error= 0.2%
It is, therefore, a good, useful and accurate interpolation.

We leave the reader with the test of the polynomial of second degree.

Here is another type of interpolation or extrapolation:

Suppose we want to calculate the value at (i,j) in red; Green values are what we rely on:

Unknown Value (i,j)=(((2 x Value (i-1,j)- Value (i-2,j))+ (2 x Value (i-1,j+1)-Value (i-2,j+2))+ (2 x Value (i,j+1)- Value (i,j+2))))/3

1,76	1,799	1,838	1,878	1,917
1,746	1,786	1,825	1,864	1,903
1,733	1,772	1,811	1,851	1,89
1,719	1,759	1,798	1,837	1,876
	1,745	1,784	1,823	1,863

The calculated value is 1.81166; an error of 0%. Very accurately, yes, but we need many values around.

Calculating the complete board:

We use for this, the same interpolation-extrapolation above:

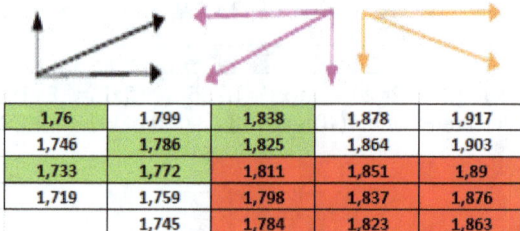

1,76	1,799	1,838	1,878	1,917
1,746	1,786	1,825	1,864	1,903
1,733	1,772	1,811	1,851	1,89
1,719	1,759	1,798	1,837	1,876
	1,745	1,784	1,823	1,863

Calculation:

1,81166667	1,85033333	1,89
1,798	1,837	1,87666667
	1,82333333	1,86233333

Error in %:

0,03681207	-0,03601657	0
0	0	0,0355366
	0,01828488	-0,03578458

To apply this procedure and obtain an "acceptable" tablewe must have known points scattered throughout the table; in case of not having, the error, as we move us away from the area where we have points or where we started, will be higher.

The function from which we generate an unknown value, has 7 unknowns; we must get a system of seven equations or schemas, to solve it; this is a linear system.

Calculating a value and complete board: Plane by 3 points:

We will generate a plane passing through 3 points in this way, interpolate and extrapolate as the case, all other values in the table to study. We will see different strategies of choice of these 3 points.
Our work table is again:

1,603	1,664	1,702	1,765	1,828	1,85	1,873	1,895	1,917
	1,702	1,737	1,769	1,8	1,818	1,835	1,853	1,871
			1,772	1,772	1,785	1,798	1,811	1,824
				1,755	1,765	1,775	1,785	1,796
					1,745	1,752	1,76	1,767

We create a plane passing through the points marked in blue:

a)

1,603	1,664	1,702	1,765	1,828	1,85	1,873	1,895	1,917
	1,702	1,737	1,769	1,8	1,818	1,835	1,853	1,871
			1,772	1,772	1,785	1,798	1,811	1,824
				1,755	1,765	1,775	1,785	1,796
					1,745	1,752	1,76	1,767

$$-0.868x + 3.14y - 160z = -241.98$$

With the plane thus generated, we calculate the complete board:

1.603	1.642	1.681	1.721	1.76	1.799	1.838	1.878	1.917
1.589	1.629	1.668	1.707	1.746	1.786	1.825	1.864	1.903
1.576	1.615	1.654	1.694	1.733	1.772	1.811	1.851	1.89
1.562	1.602	1.641	1.68	1.719	1.759	1.798	1.837	1.876
1.549	1.588	1.627	1.666	1.706	1.745	1.784	1.823	1.863

Let us see the error in each box (%):

0,000	1,340	1,249	2,557	3,864	2,835	1,904	0,905	0,000
		4,137	3,632	3,093	1,792	0,548	-0,590	-1,682
			4,604	2,250	0,734	-0,718	-2,161	-3,492
				2,094	0,341	-1,279	-2,831	-4,264
					0,000	-1,794	-3,456	-5,153

The sum of all errors is: 58

b)

1,603	1,664	1,702	1,765	1,828	1,85	1,873	1,895	1,917
	1,702	1,737	1,769	1,8	1,818	1,835	1,853	1,871
			1,772	1,772	1,785	1,798	1,811	1,824
				1,755	1,765	1,775	1,785	1,796
					1,745	1,752	1,76	1,767

$$-2.4x + 3.14y - 160z = -249.64$$

1.603	1.642	1.681	1.721	1.76	1.799	1.838	1.878	1.917
1.565	1.605	1.644	1.683	1.722	1.762	1.801	1.84	1.879
1.528	1.567	1.606	1.646	1.685	1.724	1.763	1.803	1.842
1.49	1.53	1.569	1.608	1.647	1.687	1.726	1.765	1.804
1.453	1.492	1.531	1.571	1.61	1.649	1.688	1.728	1.767

0,000	1,340	1,249	2,557	3,864	2,835	1,904	1,337	0,000
	6,044	5,657	5.110	4,530	3,178	1,888	0,707	-0,426
			7,655	5,163	3,538	1,995	0,444	-0,977
				6,557	4,624	2,839	1,133	-0,443
					5,822	3,791	1,852	0,000

The sum of all errors is: 100

c)

1,603	1,664	1,702	1,765	1,828	1,85	1,873	1,895	1,917
	1,702	1,737	1,769	1,8	1,818	1,835	1,853	1,871
			1,772	1,772	1,785	1,798	1,811	1,824
				1,755	1,765	1,775	1,785	1,796
					1,745	1,752	1,76	1,767

$$0.78x - 0.3075y + 60z = 110.745$$

-11,485	-8,672	-7,096	-4,180	-1,349	-0,698	0,000	0,637	1,268
	-4,863	-3,446	-2,265	-1,099	-0,656	-0,325	0,108	0,537
			-0,281	-0,895	-0,723	-0,553	-0,385	-0,273
				0,000	0,000	-0,056	-0,056	0,000
					0,692	0,516	0,399	0,170

The sum of all errors is: 38

Conclusions:

- The more equilateral the triangle formed by the "base points" is, the greater the accuracy.
- The more area covers the same triangle, greater accuracy.

Note:

Instead of having added errors to determine the suitability, we could add the squares of the errors.

We could try more models and fine tune the process; some options presented below:

- Calculation of a value and complete board: area by 5 points:

Ellipsoid passing through 5 points

$$\begin{vmatrix} x^2+y^2 & z^2 & x & y & z & 1 \\ x_1^2+y_1^2 & z_1^2 & x_1 & y_1 & z_1 & 1 \\ x_2^2+y_2^2 & z_2^2 & x_2 & y_2 & z_2 & 1 \\ x_3^2+y_3^2 & z_3^2 & x_3 & y_3 & z_3 & 1 \\ x_4^2+y_4^2 & z_4^2 & x_4 & y_4 & z_4 & 1 \\ x_5^2+y_5^2 & z_5^2 & x_5 & y_5 & z_5 & 1 \end{vmatrix} = 0$$

- Calculation of a value and complete board: Bilinear interpolation:

BILINEAR INTERPOLATION

$$G(x,y+1) \quad G(x+1,y+1)$$

$$y_i$$

$$G(x,y) \; x_i \; G(x+1,y)$$

$$p = x - x_i$$
$$q = y - y_i$$
$$G(x_i, y_i) = (1-p)(1-q)G(x,y) + (1-p)qG(x,y+1) + p(1-q)G(x+1,y) + pqG(x+1,y+1)$$
$$\frac{\partial G}{\partial x_i} = (1-q)(G(x+1,y) - G(x,y)) + q(G(x+1,y+1) - G(x,y+1))$$
$$\frac{\partial G}{\partial y_i} = (1-p)(G(x,y+1) - G(x,y)) + p(G(x+1,y+1) - G(x+1,y))$$

Calculation of a value and complete board: Surface by 4 points:

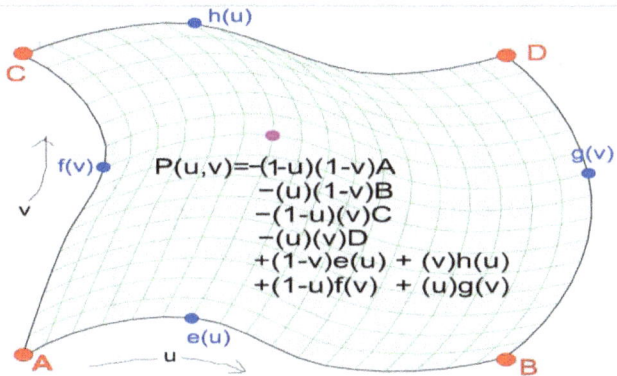

$$P(u,v) = -(1-u)(1-v)A$$
$$-(u)(1-v)B$$
$$-(1-u)(v)C$$
$$-(u)(v)D$$
$$+(1-v)e(u) + (v)h(u)$$
$$+(1-u)f(v) + (u)g(v)$$

- Calculation of a value and complete board: Bezier surface.
- Calculation of a value and complete board: Bicubic interpolation:

Either way, we will now focus on the case of a sphere: we assume that the best geometry that fits our table is a sphere and work with it.

ABSTRACT

We will establish what points of an 'aeromap' are best placed to, from them, generate spherical surfaces that represent approximately one full table of the aeromap of a vehicle. We conducted the study with the Dallara F-302 for 2002 F3.

THEORETICAL INTRODUCTION

A quadric is the geometric space that verifies a quadratic equation in three unknowns of the type:

$$A \cdot x^2 + B \cdot y^2 + C \cdot z^2 + D \cdot x \cdot y + E \cdot x \cdot z + F \cdot y \cdot z + G \cdot x + H \cdot y + I \cdot z + J = 0$$

Among the types of quadrics, we talk about sphere when the coefficients associated with x^2, y^2, z^2 are the same and the coefficients associated with the rectangular terms $x \cdot y$, $x \cdot z \cdot zyy$ are void.

The equation of a spherical surface passing through 4 points is obtained by developing the following determinant:

$$\begin{vmatrix} x^2 + y^2 + z^2 & x & y & z & 1 \\ x_1^2 + y_1^2 + z_1^2 & x_1 & y_1 & z_1 & 1 \\ x_2^2 + y_2^2 + z_2^2 & x_2 & y_2 & z_2 & 1 \\ x_3^2 + y_3^2 + z_3^2 & x_3 & y_3 & z_3 & 1 \\ x_4^2 + y_4^2 + z_4^2 & x_4 & y_4 & z_4 & 1 \end{vmatrix} = 0$$

Where specified points are:

$$(x_1, y_1, z_1), (x_2, y_2, z_2), (x_3, y_3, z_3) \text{ y } (x_4, y_4, z_4)$$

PROCEDURE

The way we're going to approach the study is taking all possible combinations of 4 points from the 32 of the range of the 'aeromap' of the vehicle and creating from them spherical surfaces. Since the order in which you choose is unimportant and points cannot be repeated, the number of combinations will be:

$$\binom{32}{4} = \frac{32!}{28! \cdot 4!} = \frac{32 \cdot 31 \cdot 30 \cdot 29}{4 \cdot 3 \cdot 2} = 35960$$

From these created surfaces we can generate an 'aeromap' to compare with that obtained in the wind tunnel test. We should note that among the possible combinations we have the ones that take all the points of the same value 'FRH' or those that do about the same value of 'RRH'. In such cases we would not get a surface but a straight in dependence on x or in y. These cases will be obviously excluded from the analysis.

We start from the next couple of 'aeromaps' corresponding to a Dallara F-302 Formula 3 in which values in the work range are highlighted in orange background. The first one corresponds to a low load configuration and the second one to a high load configuration:

Cz	6	8	10	12	14	16	18	20	22	24	26	28	30
5,0	1,603	1,664	1,702	1,765	1,828	1,850	1,873	1,895	1,917	1,924	1.932	1,939	1.946
7,5	1,665	1,702	1,737	1,769	1,800	1,818	1,835	1,853	1,871	1,873	1,875	1,877	1.880
10,0	1,793	1,778	1,772	1,772	1,772	1,785	1,798	1,811	1,824	1,821	1,819	1,816	1.813
12,5	1,798	1,773	1,766	1,760	1,755	1,765	1,775	1,785	1,796	1,795	1,794	1,793	1,793
15,0	1,806	1,774	1,760	1,749	1,738	1,745	1,752	1,760	1,767	1,768	1,770	1,771	1,772

Cz	6	8	10	12	14	16	18	20	22	24	26	28	30
5,0	2,125	2,173	2,221	2,269	2,317	2,335	2,353	2,371	2,389	2.389	2.389	2,389	2.389
7,5	2,180	2,206	2,232	2,257	2,283	2,298	2,314	2,329	2,344	2.344	2,343	2,343	2,342
10,0	2,235	2,239	2,242	2,246	2,249	2,262	2,274	2,287	2,299	2.298	2,297	2,296	2.295
12,5	2,263	2,257	2,252	2,247	2,242	2,252	2,261	2,270	2,280	2.277	2,275	2,272	2,270
15,0	2,290	2,276	2,263	2,249	2,236	2,242	2,248	2,254	2,260	2.256	2,253	2,249	2,245

The error that allows us to qualify for better or worse a spherical surfaceis defined as follows:

$$Error = \sum_{i=1}^{32} \left(\frac{C_z T_i - C_z T^{exp}{}_i}{C_z T^{exp}{}_i} \right)^2$$

Where $C_z T^{exp}{}_i$ is the i-th value of the 'aeromap' obtained experimentally and $C_z T_i$ its i-th associated value obtained as the 'z' value of the generated spherical surface.

Once done all possible spherical surfaces and calculated errors defined above, we analyze what 'aeromap' points are those that show better surfaces to represent the whole 'aeromap' in an approximate way.

RESULTS AND DISCUSSION

Then we will show and discuss individually the results obtained for each of the two 'aeromaps' of the Dallara F-302 to finish with a joint analysis that allows us to draw overall conclusions.

For the 'aeromap' corresponding to the low downforce settings we obtained the following results:

Order	Error	Boxes				Major individual error (%)
1	0,009858	6	21	22	24	5.20
2	0,009925	3	21	22	29	6.20
3	0,010009	3	20	22	29	6.22
4	0,010043	3	20	21	29	6.22
5	0,010183	3	19	22	28	6.28
6	0,010349	6	20	22	24	5.33
7	0,010581	6	21	22	28	5.15
8	0,010713	6	20	21	24	5.47
9	0,010935	6	16	17	28	4.46
10	0,010957	6	15	17	28	4.48
11	0,010958	3	21	22	24	6.53
12	0,010998	6	15	16	28	4.49
13	0,011098	6	17	19	28	4.58
14	0,011103	3	6	16	29	5.79
15	0,011119	6	20	22	28	5.40
16	0,011214	3	20	27	28	6.68

Where the 'Error' parameter we have defined in 'Procedure' of this article and the so-called 'Major individual error ' is the largest error recorded for one box of the 'aeromap'. As can be seen, the box 3 and 6 are not repeated in the same quadruple until number 14 and in all more efficient quadruples, with smaller errors, always appears one, so at least, a priori, we can consider them as privileged. With boxes 24, 28 and 29 occurs the same. Nor they are repeated in the same values quadruple but one always appears in one of the quadruples that best fit the experimental 'aeromap' obtained from the wind tunnel. Box 28 is the most often repeated of three. It appears in eight of the sixteen quadruples. Out of these two groups of cells (the one formed by 3, 6 and the one formed by 24, 28, 29) is box 22 with eight appearances the most often repeated in quadruples.

Following is the boxes, the values of 'FRH' and 'RRH', used to build the 16 spheres which best fit the 'aeromap' experimentally obtained:

	6	8	10	12	14	16	18	20	22
5	1	2	4	6	9	13	18	23	28
7.5		3	5	7	10	14	19	24	29
10				8	11	15	20	25	30
12.5					12	16	21	26	31
15						17	22	27	32

The cells involved in the generation of the 16 areas that best represent the whole 'aeromap' have been highlighted in yellow. As we can infer from the above mentioned, the cells used can be separated into three groups. On the one hand the one formed by boxes 3 and 6, on the other hand the one formed by boxes 15, 16, 17, 19, 20, 21, 22, 27 and finally the group of 24, 28 and 29. To generate a sphere with the least possible error we take a box of the first group, two boxes of the second group and one of the third group box.

Let's see what happens with the second 'aeromap': For the 'aeromap' corresponding to the high downforce configuration we obtained the following results:

Order	Error	Boxes				Major individual error (%)
1	0,003618	6	19	21	28	3.44
2	0,003911	3	15	24	27	4.08
3	0,003920	3	15	21	29	3.95
4	0,003982	6	15	21	28	3.76
5	0,004042	6	15	19	21	3.62
6	0,004046	3	6	21	24	4.09
7	0,004048	3	11	16	24	3.72
8	0,004053	6	11	22	29	4.12
9	0,004059	3	6	15	24	4.10
10	0,004066	3	15	21	24	4.10
11	0,004075	3	6	15	21	4.11
12	0,004081	6	15	21	24	4.12
13	0,004134	6	15	21	23	3.50
14	0,004181	3	6	16	19	3.79
15	0,004186	6	19	23	27	3.47
16	0,004220	3	11	22	28	3.92
17	0,004231	3	19	21	28	4.18
18	0,004245	3	19	27	28	4.21

First it should be noted as the 'aeromap' corresponding to the configuration with more downforce (on the text) is better accommodated to a sphere that the corresponding to low load. The error defined in 'procedure' is, in the best generated surface, around 2.7 times lower with the latter 'aeromap'. The 'major individual error' of the boxes are also lower for this 'aeromap', as can be seen in the tables. Said this, we can see, also for this 'aeromap', that boxes 3 and 6 are privileged: Of the 18 best spherical surfaces generated no one exclude any of these boxes. Following is the boxes, the values of 'FRH' and 'RRH', used to build the 13 spheres which best fit the 'aeromap' experimentally obtained:

	6	8	10	12	14	16	18	20	22
5	1	2	4	6	9	13	18	23	28
7.5		3	5	7	10	14	19	24	29
10				8	11	15	20	25	30
12.5					12	16	21	26	31
15						17	22	27	32

CONCLUSIONS

According to the results obtained for each of the two 'aeromaps' we can advise the use of a few boxes over others so that the surfaces that generate approach as much as possible the 'aeromap' gained from experience in the wind tunnel. For this, each cell will have to comply with the conditions we have imposed. According to the study conducted we must select the boxes taking:

One or two of the couple: 3 y 6
One or two of the couple: 24 y 28
The cell: 21

METHODS AND SYSTEMS TO MANTAIN HEIGHT LEGALLY

There are only two ways to achieve this:

- Acting on the suspension arms: suspension design
- Acting on the springs or dampers.

Years ago a system that kept the car at the same angle and position on the asphalt was allowed: it was the active suspension. This control was performed electronically. With the prohibition of this system, in 2012 Mercedes incorporated a system called "FRIC" which does exactly the same thing but controlling active suspension geometry by hydraulic methods; ie passively; It was "legal and complies with 2012 regulations":

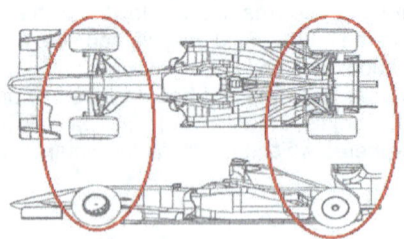

Hydraulically connecting the front and rear suspensions, so that if one changes, the other also to maintain the geometry of the car. The other advantage is that by the same method, you can control and mitigate vibrations:

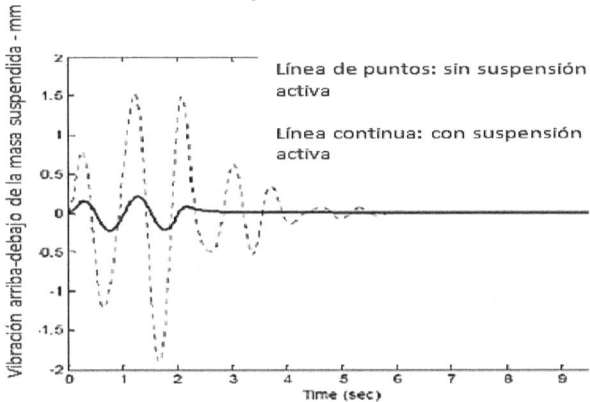

This system acts on the suspension arms, causing it to lengthen or shorten, to maintain ride height always constant; this activation is performed by pressing the brake. Another method used is to design the suspension so that the same is achieved; such designs correspond to antipitch and antidive geometries. Both geometries, respectively, are responsible for making the car does not turn up the nose when accelerating and not lower the nose when braking. That it seems that everything is wonderful and that the provision of this particular geometry solves the problem of maintaining the height relative to the tarmac; but the thing, as always, is not so simple: We must know perfectly the deformations of the tires to any solicitation; as we have seen and will see also this knowledge impacts very directly on the setup of the car and the car behavior in general, and particularly on the aerodynamics. If the car down the nose when braking, the friction coefficient of the tire increases, and therefore increases the braking capacity; in turn, in the case of cars with a lot of down force, the front spoiler produce more load which will also affect the front grip and also increase braking capacity. Therefore:

- Antipitch: optimum in straight.
- Antidive: bad at breaking.

On the other hand, if with car speed come down the rear zone, it would have a better grip so the acceleration would be greater. Conclusion: Both geometries must be conveniently used, as always.
The various effects that we can achieve about braking with this principle are:

-Front wheel:

- CT in front of wheel: *Pro-dive*
- CT behind wheel: *Anti-dive*

-Rear wheel:

- CT in front of wheel:*Pro-lift*
- CT behind wheel: *Anti-lift*

In competition usually does not exceed 30% of anti-dive (often achieved with just 1-2 degrees of inclination of the anchors) due to a combination of pros and cons elements:
→Pros of anti-dive geometry:
- Reduces the sinking of the front so that does not change the geometry of the chassis, the ground clearance and aerodynamics at heavy braking.

→ Cons of anti-dive geometry:
- Reduces the sensitivity of the suspension, so it takes a softer springs to achieve the same ratio of hardness in the wheel, which we can have problems with excessive body roll.
- Alters the perception of the driver in not perceive any sinking of the front so that can give the impression of lack of braking.

It is expressed as a% of total effect. Its value is found drawing a line between the Centre of transfer (CT) and the center of the footprint of the corresponding wheel; in the case of 2 parallel lines, it is calculated by a line parallel to the anchors from the center of the tread; then cutting this line with the vertical center of gravity "CG".
The division of the height of this point between the height of center of gravity and multiplied a hundred times give us the % effect. In the picture, the front anti-dive has a 100% effect, as the "CT" is at the same height as the CG and rear anti-crushing 50%. If, for example, we have the anchors of the suspensions in the chassis parallel and horizontal, we will have a 0% effect because the line will be parallel to the floor and 0 cm high.

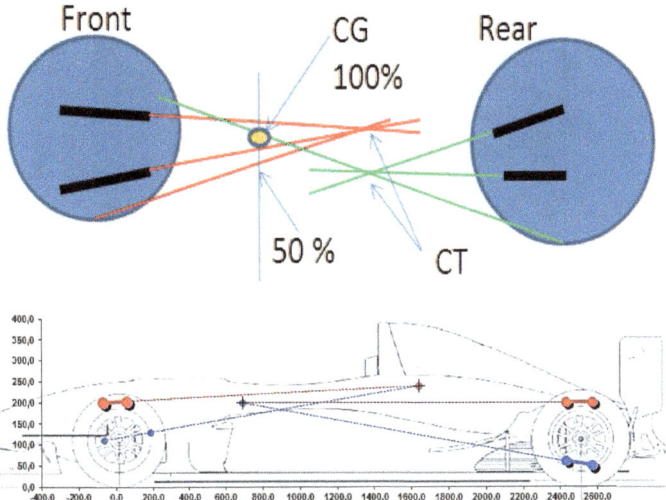

There are lot systems for that (maintain the height floor to track):

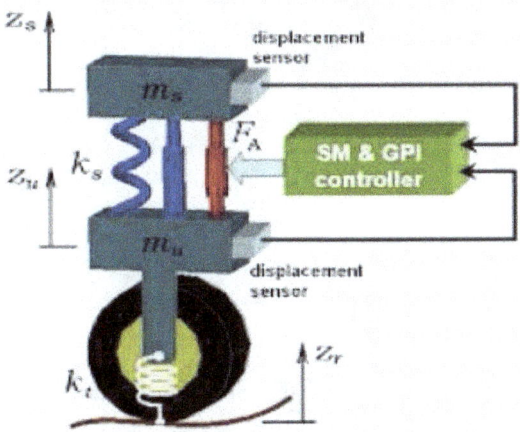

The mail is modifier the damper or spring, for maintaining the height.
Is possible for that:
- Electronic and hydraulic method:

- Aero method:

- In wheels:

➔ Resume:

The position of CDP and CG are very important:
These positions depend the:

- Over – under steering (speed, attitude, etc….).
- The grip of tire.